THE REAL ENVIRONMENTAL CRISIS

$$\int f(x)\, dx$$

$$\iiint f(x,y,z)\, dx\, dy\, dz$$

JACK M. HOLLANDER

THE REAL ENVIRONMENTAL CRISIS

*Why Poverty,
Not Affluence,
Is the Environment's
Number One Enemy*

UNIVERSITY OF CALIFORNIA PRESS
Berkeley   Los Angeles   London

University of California Press
Berkeley and Los Angeles, California

University of California Press, Ltd.
London, England

© 2003 by the Regents of the University of California
First paperback printing 2004

Library of Congress Cataloging-in-Publication Data

Hollander, Jack M.
    The real environmental crisis : why poverty, not affluence, is the
environment's number one enemy / Jack M. Hollander
        p.   cm.
    Includes bibliographical references and index.
    ISBN 0-520-24328-5 (pbk: alk. paper)
1. Poverty.   2. Sustainable development.   3. Environmental degradation—Economic aspects.   4. Environmental policy—Economic aspects.
I. Title.

HD79.E5 H648 2003
333.7—dc21   2002027171

Manufactured in the United States of America
11 10 09 08 07 06 05 04
10 9 8 7 6 5 4 3 2 1
The paper used in this publication meets the minimum requirements of
ANSI/ NISO Z39.48-1992 (R 1997) (Permanence of Paper). ∞

To Sharon, who inspires me, and so many others,

with the beauty of her music,

her intellect, and her spirit.

CONTENTS

ILLUSTRATIONS

PREFACE

The draft of this book was completed barely a month before the tragic events of September 11, 2001, thrust an entirely new set of problems and priorities onto the world stage. Although little has changed in the issues that motivated me to write the book—the impacts of poverty and affluence on the environment and the public's misunderstandings about resource and environmental issues—people's perceptions of these issues shifted dramatically, perhaps irreversibly, on September 11. Suddenly we in the affluent countries felt less insulated from the worldwide tragedy of poverty, and we now understand more clearly that poverty is a root cause, though certainly not the only cause, of the hopelessness and humiliation that beget acts of violence against fellow humans. This book makes the case that poverty is also linked to violence against the environment and that a global transition from poverty to affluence is essential to bringing about an environmentally sustainable world.

Not all environmentalists agree with this position. Some believe that the opposite is true, that the transition from poverty to affluence spells doom for the environment. Indeed, a huge gulf in worldview separates environmentalists, myself included, who are optimistic about the future of the environment from others who see only a bleak environmental future for our earth. For decades the public has been exposed mostly to the pessimistic view, a view fueled by a constant stream of bad news and doomsday predictions about resources and the environment emanating from individuals, environmental groups, and the media. No doubt, a certain level of consciousness raising by scientists and environmental groups is essential to develop and maintain people's sensitivity to environmental problems. But there is a big difference between advising caution on a slippery road and crying "fire" in a crowded theater. We've had too much of the latter, in the name of environmentalism.

Ever since the earliest days of the environmental movement, the environmental community has been deeply polarized. On one hand, many in the science and technology research community—physicists, chemists, biologists, economists, engineers, and others—enthusiastically joined new interdisciplinary groupings to study environmental problems. They were an optimistic lot and carried out the work with a strong sense of purpose and a belief that environmental problems can be solved. Some quite important new ideas for mitigation of air and water pollution, efficient use of energy, renewable energy supply sources, and technologies for clean burning of fossil fuels have been generated by these research efforts in many universities and laboratories worldwide. They contributed, and continue to contribute, significantly to a growing body of environmental knowledge and to an informed basis for government policy.

But the social and technological optimism that characterized this segment of the environmental movement was challenged by a very different kind of environmentalism, one whose philosophy embraced a strong anti-technology perspective and a distinctly pessimistic view of the future. Environmental radicalism helped catalyze the Green political parties in Europe and is now becoming a significant political force in the United States. This movement opposed nuclear electricity generation early on and, more recently, the use of traditional fossil fuels. Environmental extremism has permeated many of the world's mainline conservation organizations. The doomsday rhetoric of the environmental extreme has been willingly amplified by the media, well aware of the public's almost insatiable appetite for bad news.

Where there are strong differences in viewpoints among experts—certainly the case in environmental matters—which experts can one believe? Even among the purest of scientists there are no *pure* viewpoints, uncolored by ambitions, associations, political pressures, social pressures, financial pressures. So one finds a wide spectrum of environmental viewpoints ranging from the doomsday pessimists to the Pollyannish optimists. Most environmental professionals do not subscribe to either extreme but hold highly nuanced and contingent views of these complex subjects. Yet among the nonexpert public the dominant impression is clearly pessimistic, as the result mostly of media exaggeration. This book was written for that nonexpert public, to provide an antidote to the ubiquitous environmental exaggeration and to argue that extreme pessimism about the

Figure 1.   Cartoon from *That's Life*, by Mike Twohy.

environment is not justified by science, by economics, by demographics, or by history.

Psychologists tell us that nonexperts, in pondering which experts to believe, tend to regard the credibility of the messenger to be as important as the message itself, perhaps even more important. Whatever credibility I can claim in writing of these issues comes from three decades of involvement in environmental science and policy, mostly in academic settings. These varied experiences have given me the opportunity to participate in and to observe countless discussions and reviews of environment and resource issues, with little encumbrance by institutional or corporate allegiances. They also led to my concern about the increasing influence of environmental pessimism, and to my decision to attempt the presentation of a more balanced and optimistic picture in this book.

Although I bear sole responsibility for the content of this book, including whatever errors I may have committed, many colleagues and friends have helped me by reading and offering comments on the whole or parts of the manuscript. My thanks go to Bruce Ames, Harvey Brooks, Duncan Brown, Sydney Cameron, Joel Darmstadter, Freeman Dyson, Allan Hollander, Michael Lederer, Richard Lindzen, Sharon Mann, Richard Muller, Tihomir Novakov, John Rasmussen, Bertram Raven, Fred Singer, and Marci Li Wong. I owe a special debt to my editor at the University of California Press, Doris Kretschmer, for her wise guidance throughout the preparation of the manuscript. And I thank Mike Twohy for permission to reproduce his cartoon, which so well captures one of the book's main themes.

INTRODUCTION: *A Crisis of Pessimism*

Can you remember a day when you opened your morning newspaper *without* finding a dramatic and disturbing story about some environmental crisis that's either here already or lurks just around the corner? That would be a rare day. On one day the story may be about global warming; on the next it may be about overpopulation or air pollution or resource depletion or species extinction or sea-level rise or nuclear waste or toxic substances in our food and water. Especially jarring is the implication in most of these stories that *you and I are the enemy*—that our affluent lifestyles are chiefly responsible for upsetting nature's balance; polluting our cities, skies, and oceans; and squandering the natural resources that sustain us. Unless we change our thoughtless and wasteful ways, we are reminded, the earth will become a very inhospitable place for ourselves and our progeny.

Such media reportage reflects the pervasive pessimism about the future that has become the hallmark of today's environmental orthodoxy. Its central theme is that the *affluent* society, by its very nature, is the *polluting* society—the richer we become, the more we consume the earth's scarce resources, the more we overcrowd the planet, the more we pollute the earth's precious land, air, and water. The clear implication of this viewpoint is that the earth was a better place before humans were around to despoil it.

Some people, even some environmental scientists, genuinely subscribe to this gloomy picture of the earth's future. I do not hold that they are necessarily uninformed, or naive, or unprofessional, or captive to special interests. But they are indeed pessimistic. I am more optimistic about the earth's environmental future, and I believe there is plenty of evidence to support an optimistic, though not cornucopian, view of the environmental future. This book presents such an optimistic perspective.

In my judgment, people are not the enemy of the environment. Nor is affluence the enemy. Affluence does not inevitably foster environmental degradation. Rather, affluence fosters *environmentalism.* As people become more affluent, most become increasingly sensitive to the health and beauty of their environment. And gaining affluence helps provide the economic means to protect and enhance the environment. Of course, affluence alone does not guarantee a better environment. A sense of social responsibility is also required. Political will is also required. *But affluence is a key ingredient for ensuring a livable and sustainable environment for the future.*[1]

The real enemy of the environment is *poverty*—the tragedy of billions of the world's inhabitants who face hunger, disease, and ignorance each day of their lives. Poverty is the environmental villain; poor people are its victims. Impoverished people often do plunder their resources, pollute their environment, and overcrowd their habitats. They do these things not out of willful neglect but only out of the need to survive. They are well aware of the environmental amenities that affluent people enjoy, but they also know that for them the journey to a better environment will be long and that their immediate goal must be to escape from the clutches of poverty. They cannot navigate this long journey without assistance—assistance from generous institutions, nations, and individuals and from sincere and effective policies of their own governments.

For the affluent nations to assist people in the developing world is socially responsible and morally right. But from an environmental perspective the issue is more than ethical. It is pragmatic as well, since the environmental self-interests of the affluent would be well served by the eradication of poverty. This idea disturbs those who fear that people emerging from poverty will inevitably become "wasteful" consumers like ourselves and will only exacerbate the globe's environmental damage as they pursue the trappings of the good life. The fear is understandable, but the conclusion is wrong. Without doubt, people tasting affluence will embrace consumerism and become proud owners of property, vehicles, computers, cell phones, and the like. But they will also pursue education, good health, and leisure for themselves and their families. And *they will become environmentalists.*

Environmentalists are made, not born. In the industrial countries environmentalism arose as a reaction to the negative impacts of early industrialization and economic growth. On the way from subsistence to affluence, people developed a greater sense of social responsibility and had more time and energy to reflect about environmental quality. They had experienced environmental deterioration firsthand, and they demanded improvement. One of the great success stories of the recent half-century is, in fact, the remarkable progress the industrial societies have made, during a period of robust economic growth, in reversing the negative environmental impacts of industrialization. In the United States the air is cleaner and the drinking water purer than at any time in five decades; the food supply is more abundant and safer than ever before; the forested area is the highest in three hundred years; most rivers and lakes are clean again; and, largely because of technological innovation and the information revolution, industry, buildings, and transportation systems are more energy- and resource-efficient than at any time in the past. This is not to say that the resource/environment situation in the United States is near perfect or even totally satisfactory—of course it is not. Much more needs to be done. But undeniably, the improvements have been remarkable. They have come about in a variety of ways—through government regulation, through taxation, through financial incentives, through community actions. Most important, these environmental improvements cannot be credited solely to government, environmental organizations, or lobbyists, though each has played an important role. Rather, they have come about because the majority of citizens in this and every other democratic affluent society demands a clean and livable environment. Does this imply that the affluent have achieved an improved environment in their own lands by exporting their pollution to the lands of the poor? That has rarely been the case. (See the discussion of exporting pollution in Chapter 7.)

As the industrial societies continue to make steady progress in reclaiming their environment, they are now laying the foundation for a postindustrial future that is globally sustainable. Some elements of this foundation already exist everywhere—people's technological ingenuity, creativity in finding solutions to emerging problems, political will to get the job done. Other elements of this foundation do not yet exist or are weak. The central argument of this book is that *the essential prerequisites for a sustainable environmental future are a global transition from poverty to affluence, coupled with a transition to freedom and democracy.* Although evidence in support of this argument could be organized in a variety of ways, I have chosen to do so in the context of specific resource and environmental issues of major importance. It hardly needs saying that any argument about the future is cloaked in uncertainty, and my arguments in this book are no exception. Yet they will have served a useful purpose if they add to public understanding of the poverty–environment connection, as well as contributing to the lively and purposeful debate among environmentalists about the issues covered in the book.

My optimism about the environmental future is at odds with the environmental orthodoxy as practiced by most environmental organizations and the media, and especially reflected in the increasing stridency of their doomsday predictions of the environmental future. There is a double irony here. First, so bleak an outlook has arisen during the very period in which the affluent societies have been making their greatest environmental and economic gains; and second, the citizenry in the affluent countries overwhelmingly support a clean environment and are becoming increasingly alienated by the hyperbolic excesses committed in the name of environmentalism. Although the root causes of today's environmental pessimism are complex and intertwined with other social issues, some of the major contributing factors, as well as the paradoxes, are illuminated by a glimpse at the environmental history of the United States.

THE BIRTH OF ENVIRONMENTALISM

In its early years, the United States retained the continent's historically agrarian character, with a largely pastoral and wooded landscape from "sea to shining sea." By the mid-nineteenth century industrialization was sweeping the country, and a growing population, mostly recent immigrants, was enjoying unprecedented economic opportunities provided by the new manufacturing culture. But along with the gains from industrialization, people living and working in nineteenth-century urban areas of the United States and Britain were also experiencing signs of environmental deterioration. Cities were becoming overcrowded, skies and rivers were becoming polluted, and urban dwellers increasingly faced the twin killers of respiratory and intestinal diseases from air and water pollution.

Yet it was rural, not urban, pollution that stimulated the awakening of an American environmental movement. The first American "environmentalists" were an elite group of amateur naturalists who were disturbed by the changes to the pristine rural environment accompanying the country's industrial development—leveling of forests, overrunning of open spaces, invading of wilderness areas. Among the most idealistic of these naturalists was John Muir, who worked tirelessly for the total preservation of wilderness areas and old forests, mostly in the mountainous areas of the far West, with the hope that future generations would be able to experience the grandeur of these precious natural resources just as he experienced them. The first head of the Sierra Club (1892), Muir has rightly been called "the father of the national park system." Equally dedicated but often at loggerheads with Muir was America's first professional forester, Gifford Pinchot, who believed not in hands-off preservation but in the sustainable use of natural resources through wise management. Becoming the leader of the utilitarian wing of the conservation movement, Pinchot was appointed the first head of the U.S. Forest Service (1905) by President Theodore Roosevelt. Roosevelt was a strong and consistent ally of the conservationists, though his dedication to preserving the habitats of wild animals was due at least partly to his passion for hunting them. Drawing on the leadership of such individuals, some of the world's foremost environmental organizations, including the Sierra Club and the World Wildlife Fund, were formed, and they played a critical role during those early decades in winning public support for nature conservation.

In contrast to their early sensitivity about the rural environment, Americans generally tolerated urban pollution for another half-century. Not only was urban pollution initially perceived as an inevitable by-product of industrial production, but in the twentieth century's first two decades pollution became a symbol, at least among the working classes, of growing prosperity and an abundance of jobs. And during the Great Depression years of the 1930s, when massive unemployment returned and poverty became a fact of life for millions of Americans, chimney smoke and soot from still-operating industries became an even more welcome urban sight. Smoke in the air meant food on the table, at least for those who had jobs.

With the coming of World War II, the economic situation abruptly improved, but the environment did not. The wartime economy generated enormous production increases, full employment, and even higher levels of air and water pollution. After the war, the return to peacetime production brought an unprecedented surge of affluence and a seemingly insatiable demand for homes, automobiles, and other consumer products that had been unavailable in wartime. The pollution, unfortunately, only worsened.

But soon another kind of demand was stirring. Along with the new affluence and consumer demand, a heightened level of environmental awareness gradually evolved among the general public. This had no precedent in the earlier conservation movement, which was largely confined to a rural elite. The burgeoning postwar American middle class wanted their cities and neighborhoods to reflect their new affluence, to be attractive and healthy places to live. By the 1950s high levels of urban pollution that had been tolerated before and during the war became unacceptable to more and more Americans. By then it was no longer a laughing matter when the Cuyahoga River in Cleveland burst into flames because its surface was covered with industrial debris and slime. Or when the skies over Los Angeles became so smoggy that one could "see" the air but not the ground. Or when residents of an upstate New York community discovered that their homes had been knowingly built over an old industrial waste dump and were being threatened by leakage of toxic materials. The desire to find environmental quality at an affordable price was in fact one of the main stimuli for the exodus of millions of Americans from decaying core cities to the newly developing, still pristine suburbs.

All over the country, people began demanding cleaner air, water, and land. By the start of the 1970s both federal and state governments responded to the public's voice by creating new executive agencies dedicated to environmental protection.[2] A stream of environmental mandates and regulations soon emanated from these agencies and the legislatures, beginning a trend toward ever tighter environmental controls that continues to this day. Also proliferating during this period were nongovernmental organizations (NGOs) that focused on environmental issues, such as the Natural Resources Defense Council and the Environmental Defense Fund, which collectively soon constituted a powerful political force. These NGOs were influential in stimulating, often through legal actions, many government policies and regulations that were to play an essential role in reducing pollution. It is important to keep in mind that these environmental responses were not forced on people. Overwhelmingly, Americans have supported both government regulations and private initiatives to improve the environment. And organized environmental activism was by no means confined to the United States. Similar activities and initiatives were occurring in all the industrial countries of the noncommunist world, as a result of which thousands of environmental interest groups and NGOs function throughout the world today.

ENVIRONMENTAL SCIENCE

Besides public awareness, other developments occurred in the 1960s and 1970s that were to have profound effects on the young environmental movement. Important among these was the growing role of science. The new environmental sciences brought about a major change in the way people thought about environmental problems, shifting their focus from large and visible entities to extremely small and invisible entities. Previously, in the movement's early decades, public attention had been drawn mostly to nature's grandest creations—oceans, mountains, forests, lakes. One did not need scientific training to experience the beauty and grandeur of these natural wonders, and most anyone could also recognize the unsightliness of oil-covered lakes, smog-filled skies, and logging-disfigured forests. Earlier, such unsightliness had been perceived only as assaults on esthetic sensitivities, not as threats to health. That was to change as environmental science soon pointed to potential connections between pollution and risks to health.

Advances in analytical techniques allowed environmental chemists to detect minuscule amounts of foreign substances in air, water, and food, down to the parts-per-million or even parts-per-billion level. Such tiny concentrations usually cannot be seen, tasted, or otherwise perceived directly. Although some trace-level contaminants were introduced by newly developed industrial processes and chemicals, many trace-level substances have always been present in food and the environment as the result of natural processes. Although most environmental chemists were appropriately circumspect in describing their findings, environmental writers and the media increasingly sensationalized the issue of trace contaminants, labeling them as "toxins" whatever their amount or origin and drawing alarming connections between trace pollutants and a variety of adverse human health conditions and diseases. In most cases little or no credible evidence has been found linking trace contaminants to adverse health effects at the very low doses typically encountered,[3] yet these connections have become an indelible part of the public's environmental consciousness and fears.

During this period, environmental scientists generally enjoyed considerable public confidence, and many became influential in the budding environmental movement. A prime example of this influence was biologist Rachel Carson's enormously popular book *Silent Spring*, eloquently warning of potential harm to humans and animals from trace residues of the pesticide DDT.[4] Although published in 1962, *Silent Spring* remains a leading icon of the contemporary environmental movement.

In the years following World War II, prior to Carson's criticism of pesticide use, the pesticide DDT had been widely used in the industrial countries and to a lesser extent in developing countries. In 1970 a report by the U.S. National Academy of Sciences stated, "To only a few chemicals does man owe as great a debt as to DDT. . . . In little more than two decades, DDT has prevented five hundred million human deaths, due to malaria, that otherwise would have been inevitable."[5] So great was the influence of *Silent Spring*, however, that the use of DDT in the United States was banned by the Environmental Protection Agency in 1972,[6] and similar bans were invoked in other industrial countries. Since then there have been continuing efforts by environmental groups to extend the ban of DDT to developing countries. Such a ban would expose hundreds of millions of people, especially children, to grave risks of illness and death from malaria. Because of the interventions of many scientists, however, these efforts have thus far not been successful.[7]

Other claims made by Carson have been controversial, as well. For example, her claim that DDT is a human carcinogen has not been substantiated.[8] Some scientists also disputed her claims that DDT caused thinning in bird-egg shells and population declines in brown pelicans, bald eagles, and peregrine falcons.[9] Observers documented that the great peregrine decline in the eastern United States occurred long before any DDT was present in the environment,[10] and a British study concluded that "there is no close correlation between the decline in population of predatory birds, particularly the peregrine falcon and the sparrow hawk, and the use of DDT."[11]

THE ENVIRONMENTAL LEGACY OF VIETNAM

Although the influence of science on the environmental movement remained strong during the 1960s and 1970s, the influence of politics became even stronger. This was the era of the Vietnam War, a time when distrust of government, always endemic in the American psyche, reached new heights.[12] In this period, during which the environmental mantra "small is beautiful" became popular,[13] people's distrust extended beyond government to almost all large institutions. In particular, major technology corporations were increasingly perceived as remote and unresponsive, essentially enemies of the people. During the so-called energy crisis of the 1970s the distrust was directed especially against the major oil companies, which the media portrayed as largely responsible for the gasoline shortages accompanying the 1973 Arab oil producers' boycott.[14] Another target of distrust was the large electric utilities, which at that time were heavily engaged in constructing power plants, including nuclear power plants, to meet the nation's rapidly growing use of electricity.

A major victim of the public's loss of trust was the institution of science and technology itself. In the years following World War II, Americans had generally viewed science and scientists with awe because of the crucial roles they had played in the Allied victory (for example, the development of radar, which played a key role in Great Britain's survival in 1940, and the atomic bomb, which brought about an early end to the Pacific War in 1945). As a result, U.S. scientists were blessed with unprecedented increases in government support for their research during the 1950s and early 1960s. But awe gave way to distrust during the Vietnam period. A prime target of this enmity was the scientific establishment generally but particularly the nuclear power establishment, which in that day came to symbolize the perceived excesses of science and technology. An example of this distrust was the 1979 hit film *The China Syndrome*, portraying nuclear industry executives as villains responsible for a fictional nuclear reactor accident with mass fatalities. Almost coincident with the release of this film, the accident at the Three Mile Island nuclear plant happened; despite hysterical media reporting, no injuries or fatalities actually resulted.

It is somewhat paradoxical that the public's confidence in environmental science grew rapidly during the 1960s and 1970s, a period during which the environmental scientists were bringing mostly bad news yet during which confidence in the larger scientific establishment eroded rapidly, even though science and technology were continuing to enhance the quality of people's lives. The public's growing antipathy to the Vietnam War and technology's role in that conflict were probably major factors in creating this anomaly.

TRANSFORMATION TO PESSIMISM

The Vietnam period also saw the beginnings of change in the image of environmentalism, from champion of nature's grandeur and source of optimism and vision to its current sense of doom and gloom about the earth's future. In the new environmental politics, "pro-environment" has become increasingly identified with anti-technology attitudes and, especially, with antinuclear politics. Starting in Europe, opposition to nuclear-generated electricity has long been a principal plank in the platforms of the Green political parties. And the U.S. Green Party's 2000 platform called for "early retirement of nuclear power reactors"; a national shift away from "corporate industrial farming," which it labeled as "biodevastation"; and rejection of agreements encouraging trade liberalization, such as the World Trade Organization, which it portrays as "run by corporate interests unaccountable to public input or even legal challenge."[15]

The media have played a major role in encouraging the growth of environmental pessimism and technophobia by focusing on worst-case, doomsday scenarios in reporting environmental subjects and consistently underplaying the remarkable progress being made by the affluent societies in enhancing the quality of the environment.

The real enemies of environmental progress are poverty and tyranny, not technology or global markets. On the contrary, technological innovation enabled by affluence and freedom has been a major source of the environmental progress already made by the industrial societies, and the global penetration of innovative technologies will most likely be a crucial ingredient for achieving a future sustainable environment throughout the world. Unfortunately, the reality of environmental progress and promise is obscured by the doomsday rhetoric propounded in recent years by many environmental groups and amplified by the media. Here are a few examples:

In a 1998 advertisement, the respected World Wildlife Fund tells us that "forests are being cleared. Oceans overfished. Toxic chemicals are everywhere. Not just individual plants and animals, but entire ecosystems are in danger of disappearing forever. And we will all suffer from these losses. Fewer than 500 days remain in this century, and the fate of the planet rests on choices we make today" (full-page advertisement in *New York Times,* August 21, 1998).

And the venerable Sierra Club claims that "the human race is engaged in the largest and most dangerous experiment in history—an experiment to see what will happen to our health and the health of the planet when we change our atmosphere and our climate.... The rapid buildup of carbon dioxide and other greenhouse gases in our atmosphere is the source of the problem. By burning ever increasing quantities of coal, oil and gas we are choking our planet in a cloud of this pollution. If we don't begin to act now to curb global warming, our children will live in a world where the climate will be far less hospitable than it is today" (Sierra Club global warming Internet web site, www.sierraclub.org/globalwarming, March 1999).

The Union of Concerned Scientists (UCS) warns "all humanity of what lies ahead. A great change in our stewardship of the earth and the life on it is required if vast human misery is to be avoided and our global home on this planet is not be irretrievably mutilated. The environment is suffering critical stress.... Our massive tampering with the world's interdependent web of life, coupled with the environmental damage inflicted by deforestation, species loss and climate change, could trigger widespread adverse effects, including unpredictable collapses of critical biological systems whose interactions and dynamics we only imperfectly understand.... The earth is finite. Its ability to absorb wastes and destructive effluent is finite. Its ability to provide food and energy is finite. Its ability to provide for growing numbers of people is finite. And we are fast approaching many of the earth's limits. No more than one or a few decades remain before the chance to avert the threats we now confront will be lost and the prospects for humanity immeasurably diminished" ("World Scientists' Warning to Humanity," issued by the UCS on November 18, 1992, available at www.ucsusa.org/about/warning.html).

Typical of today's environmental pessimism, these doomsday pronouncements contain grains of truth embedded in a sea of exaggeration. Without jumping ahead into the details of the scientific subjects they encompass, which is the task of subsequent chapters, I assert here that such broad-brush statements mislead the public and, in some instances, are scientifically inaccurate. For example, they usually represent environmental quality as rapidly deteriorating, which is not the case. They usually represent the earth's productive capacity as rapidly diminishing, which is not the case. They usually represent population growth as a global threat, which is not the case. And they usually represent global warming as definitely linked to human activities, which has not been established. Countering such environmental pessimism with a factual basis for environmental optimism is one of the objectives of this book.

OPTIMISM, NOT INACTION

Please do not misunderstand me. Espousing optimism about the environment does not imply being complacent or sweeping environmental problems under the rug. On the contrary, optimism implies a "can do" attitude that makes success in dealing with such problems more likely. Despair and inaction are more likely to arise from pessimism about the future than from optimism. Nor does environmental optimism equate with denial. Of course, real environmental concerns are still with us. They always have been, and they always will be. As long as humans, imperfect species that we are, live together in this increasingly interdependent global village, there will be problems arising from people's activities and interactions, as well as risks arising from human adventures and technological innovations. The environment is no exception. Although, obviously, not all environmental problems are caused by human activities, humans everywhere bear a collective responsibility to care for this planet as best we can, on the basis of the scientific knowledge we have.

Without question, environmental organizations and the media have played a historically important role in bringing important information about the environment to public attention. They should continue to do so. But performing the role of environmental watchdog does not confer license to exaggerate, mislead, or strike fear in the hearts of a largely supportive public earnestly looking for information and guidance. Scientists, specialist organizations (whether representing environmental or other interests), and the media have a collective responsibility not to cross the line separating truth, however well or poorly known, from self-serving rhetoric. Unfortunately, by exaggerating many environmental problems far out of proportion to the actual or potential threats they may pose to society's future, the purveyors of doomsday rhetoric create a climate of confusion and fear about the environment among a citizenry inadequately equipped with the scientific background needed to calibrate such rhetoric.

How could people not become fearful about global warming, for example, when they are bombarded incessantly with alarming and simplistic predictions of global catastrophe from climate change that is purportedly caused by human activities? In truth, climate change is a dynamic natural phenomenon that has been occurring ever since the earth was formed millions of years ago, and the extent of human culpability for perturbing this natural system is far from established. Climate science is so extraordinarily complex that not even leading climate scientists profess to understand climate change fully. One thing that climate scientists do understand, however, is that current predictions of future climate are based almost entirely on computer simulations. Although simulations are a widely used tool in science research generally and are essential for meteorologists' short-term weather predictions, they do not provide an adequate basis for the catastrophic generalizations about future climate often made by environmental organizations and the media. In any case, for most of us it is difficult to distinguish between solid empirical evidence and speculation based on highly uncertain computer models.

Environmental exaggeration also emanates on occasion from political leaders. For example, in his book *Earth in the Balance*, former vice-president Al Gore states that climate change is "the most serious threat we've ever faced," and "Our insatiable drive to rummage deep beneath the surface of the earth, remove all of the coal, petroleum, and other fossil fuels we can find, then burn them as quickly as they are found—in the process filling the atmosphere with carbon dioxide and other pollutants—is a willful expansion of our dysfunctional civilization into vulnerable parts of the natural world."[16] In contrast to the book's extreme rhetoric, Gore's actual voting record on environmental issues in the Senate was centrist.[17]

With environmental matters, as with most others, informed discussion is the key to effective decision making in a democratic society. Extreme rhetoric serves less to catalyze rational discussion of issues than it does to polarize people's views and create fear and confusion about the environment. Some scientists argue (usually in private[18]) that creating fears about environmental risks is an effective antidote to public apathy and complacency and that the public's environmental fears can take credit for much of their support for environmental actions. I take issue with that view and prefer to believe that a truthfully informed public is more likely than a fearful public to be supportive of meaningful responses. I would place my bets that the wisest public choices about the environment will come about from disciplined presentations by scientists, and others, of research results and from contending interpretations unembellished by exaggerations and doomsday scenarios.

When individuals and the media in the affluent countries characterize as imminent threats such issues as overpopulation, resource exhaustion, and global warming, they cause more than fear: they cause actual harm by diverting people's attention and, more important, their resources from critical global problems that cry out for solution, especially the proliferation of weapons of mass destruction and the world's most formidable and pervasive environmental problem—poverty.

ENVIRONMENT OF THE POOR

People living in poverty perceive the environment very differently from the affluent. To the world's poor—several billion people—the principal environmental problems are local, not global. They are not the stuff of media headlines or complicated scientific theories. They are mundane, pervasive, and painfully obvious:

- HUNGER—chronic undernourishment of a billion children and adults caused not only by scarcity of food resources but by poverty, war, and government tyranny and incompetence.

- CONTAMINATED WATER SUPPLIES—a major cause of chronic disease and mortality in the third world.

- DISEASES—rampant in the poorest countries. Most could be readily eradicated by modern medicine, while others, including the AIDS epidemic in Africa, could be mitigated by effective public health programs and drug treatments available to the affluent.

· SCARCITY—insufficient local supplies of fuelwood and other resources, owing not to intrinsic scarcity but to generations of overexploitation and underreplenishment as part of the constant struggle for survival.

· LACK OF EDUCATION AND SOCIAL INEQUALITY, ESPECIALLY OF WOMEN —lack of education resulting in high birthrates and increasing the difficulty for families to escape from the dungeons of poverty.

Although these deplorable environmental conditions can be attributed partly to poverty itself, the governments of many poor countries must share responsibility. Many government development policies have been conceived out of selfishness, incompetence, or maliciousness, and some have either failed to help the poor or even worsened their plight. And the very resources upon which the poor depend have in some cases been plundered through corrupt government policies. Worse yet, the constant scourge of wars between and within the world's poorest nations, as well as between rich and poor nations, has enormously exacerbated the inherent ills of poverty.

The challenges for overcoming global poverty are immense and cannot be overstated. How then can this writer be optimistic about the environmental future, given that poverty and a degraded environment are so inextricably intertwined? My optimism arises from several strongly held convictions.

First, my conviction that there is an absolute human obligation, increasingly recognized by people everywhere, that the world must lift its poor out of poverty. In spite of the ubiquitous forces of selfishness, ignorance, and tyranny working to perpetuate poverty and inequality, progress is being made—halting and slow but real nonetheless. In developing countries, a child born today can expect to live eight years longer than one born thirty years ago. Five times more rural families have access to safe water, and average incomes have almost doubled.

Second, my conviction that the vicious and self-perpetuating cycle that connects poverty and environmental degradation can best be broken by attacking and eliminating the source of the problem—poverty.

Third, my conviction, based on history and science, that affluence and freedom are friends to the environment, indeed, that the road to affluence and freedom provides the only practical pathway to achieving a sustainable future environment.

These convictions provide the motivation and intellectual foundation for this book.

With history as our guide, we can be confident that today's poor peoples, as they begin climbing the economic ladder and enjoying some measures of freedom, will attend first to basic personal and family problems of sustenance and health, just as yesterday's poor did. With the increase of freedom and affluence—both are crucial—people are then likely to become motivated and increasingly able to apply the necessary political will, economic resources, and technological ingenuity to address environmental issues more broadly.[19]

Despite much rhetoric to the contrary, there is no inherent conflict between a healthy economy and environmental quality; actually they go hand in hand. Is it not persuasive that for decades the robust economic growth of the affluent societies has coincided with their continuing environmental improvement? For the future, a major key to environmental quality, for both the emerging and industrial economies, will be development and use of innovative technologies that are *both* economically attractive and environmentally friendly. Fortunately, today's developing societies hold a tremendous advantage over yesterday's. They do not need to tread through the entire learning experience in each technology area; instead they can "leapfrog" over the pathways (and mistakes) of the industrial pioneers and jump straightaway to the environmentally kinder and smarter technologies of the twenty-first century.

There is also little basis for the fear that worldwide economic development will bring about massive environmental deterioration from the newly affluent becoming unrestrained consumers imitating the technology-oriented ways of the rich. In this century consumerism can increasingly mean replacing old and polluting technologies with new, resource-efficient and environmentally friendly technologies. Technological innovation and economic efficiency—the major keys to environmental quality—can be expected to take root increasingly in the developing nations as they make the transition to democracy and affluence. Supported by new technologies and management arrangements, agriculture, fishing, and manufacturing in the developing world have the potential eventually to become resource efficient and environmentally sustainable. As our knowledge increases, an increasing awareness of the importance of healthy ecosystems—a critical factor to achieving a sustainable environment—can be expected to develop among people everywhere. Gradually, both the poor and the rich will reduce the unwise use of forests and other natural resources, as all people progress toward affluence and democratic choice.

Nor is the fear justified that development will bring with it unsustainable exploitation of energy resources. Although it is clear that economic growth will bring about substantial increases in demand for energy *services* (such as transportation, heating, lighting, and information processing), the growth in actual energy-resource consumption can be considerably reduced by efficiency gains of the technologies supplying both energy and energy services. (For example, compact fluorescent lightbulbs, still in their infancy in terms of technical development and consumer acceptance, use only a quarter as much electricity as standard incandescent bulbs.) The amounts of fossil fuels consumed will continue to increase for several decades because of technological inertia, but in the longer term cleaner and more efficient energy technologies will become economically accessible in the developing world, and these have the potential to reduce greatly the pollution problems traditionally associated with fossil fuel burning. Another example: millions more vehicles will be on the roads in the developing countries, but they will be tomorrow's high-tech low polluters rather than yesterday's low-tech high polluters.

SUSTAINABILITY WITH AFFLUENCE

The core message of this book is that an environmentally sustainable future is within reach for the entire world provided that affluence and democracy replace poverty and tyranny as the dominant human condition. People who have the means to support investments in a healthy environment, and the freedom to do so, can be trusted to make wise environmental choices provided they are honestly informed about the costs and benefits of available options in relation to other social choices that they constantly make. But in a democracy all sides must be heard. Unfortunately it is difficult today for voices of environmental optimism to be heard over the cacophony of pessimism and fear mongering emanating from some environmental groups and the media. In the name of environmentalism, their pessimistic and divisive exaggerations have become increasingly alienating and may even be counterproductive to the achievement of long-term environmental goals. Many thoughtful citizens in the industrial countries, genuinely supportive of environmental quality but bewildered about the actual state of the environment, have grown suspicious of all environmental politics, whether emanating from the left or the right, and now increasingly distrust the disparate pronouncements even of environmental experts. Equally disturbing, policy makers in the international donor community increasingly turn away from important science-based projects—for example, research in genetically modified agricultural products—for fear of antagonizing powerful environmental lobbying groups.[20]

Whereas there were once grounds for confidence that the self-interests of environmental groups coincided with the public interest, today the exaggerations and doomsaying can be seen as self-serving marketing devices, in the same way that the public-relations exaggerations of private industry are understood as marketing devices. In order for that confidence to be restored, the environmental rhetoric needs to be muted, the political polarization needs to be diminished, and civility needs to be restored to the environmental dialogue. The public, overwhelmingly supportive of environmental goals, has the right to expect the highest standards of integrity from its environmental representatives—whether in government, industry, academia, or interest groups—in defining and explaining the world's environmental challenges.

This book argues that optimism about the environmental future is warranted by what we do know, even though there is much that we do not know. This optimism is based partly on the historical record of environmental improvement and current research, but even more, it recognizes the promise of sustained technological innovation catalyzed by human ingenuity in an increasingly affluent and democratic world.

Today, as part of the natural forces of history, the world is continuing its march toward a global society. Globalization will play a major role in bringing increased affluence and democratic choice to billions of people. The core issues of this book are not about globalization or the global economy, for example, questions relating to the comparative incomes and working conditions of workers in the developing countries today. I take it as a given that in this century family incomes in most of the developing world will continue to move upward, as they are now doing,[21] even though the rate of improvement in particular times and places will appear slow and erratic.

The core debate is about the effects of affluence on the environment. The debate can be framed around my proposition that affluence promotes true environmentalism, versus the orthodox view that affluence promotes a mindless consumerism that irreparably damages the environment. Obviously, neither proposition can be scientifically "proved" since each refers to the future, but the preponderance of evidence favors the notion of a positive link between affluence and environmental quality. And the evidence also shows that we are not dealing here with a global zero-sum game, where environmental improvement in one place (rich countries) would mean a deterioration in another place (poor countries).

In the following chapters, evidence bearing on the nature of the affluence–environment link is presented and analyzed. For the most part the discussions focus on individual environmental and resource issues that are recognized as critically important to the attainment of a sustainable environment. The major issues are explored in the context of the following supporting themes:

- Poverty is the world's most critical environmental problem. Reducing poverty throughout the world should be a top priority for environmentalists. Human development should include not only freedom of economic choices but also freedom of democratic choices.

- Affluence and the technological innovation it enables are among the most important ingredients for achieving a future sustainable global environment.

1

A WORLD APART

Nearly everyone cares about the environment. But what exactly is "the environment"? That depends on how and where you live. If you are an American, you may occasionally ponder the media's claims that last year's hot summer was a precursor of catastrophic global warming, but in any case you probably perceive such environmental scenarios as somewhat esoteric and remote from your daily life. If you are a welder in a Chinese bicycle factory, in contrast, you are fully aware of the serious water and air pollution that China's rapid industrialization has brought to your region, but you probably accept the pollution with forbearance because the bicycle factory provides a steady job that enables you to support your family. Yet if you are a subsistence farmer in sub-Saharan Africa living on the brink of starvation, you probably think of the environment as nature's fickle preserve—the land and animals that in good years barely keep you and your family alive and in bad years bring starvation and disease. The environment of the rich and the environment of the poor are indeed a world apart.

Life on the brink of starvation has in fact been the fate of the vast majority of humans throughout history. To people living in such poverty, the environment has always had only one meaning and purpose: it is the source of the food and shelter needed to survive and reproduce. Yet even at the start of the twenty-first century, the most affluent ever, the environment of the poor still does not provide sufficient food for them. Their hunger is not a transitory condition—it is chronic, debilitating, and deadly, blighting the lives of all who are affected.

Approximately one billion people—one in every six people on earth—do not have enough to eat. Almost two-thirds of these chronically undernourished people (525 million) live in Asia and the Pacific. India alone has

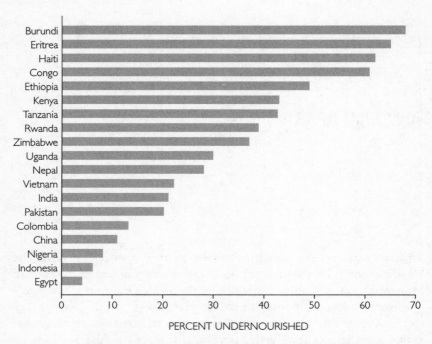

Figure 2. Number of undernourished people in selected countries (1999). Data are from United Nations Food and Agricultural Organization, *The State of Food Insecurity in the World*.

more undernourished (204 million) than all of sub-Saharan Africa, where 180 million go hungry. China is a close third, with 164 million hungry people.[1] Every year over 6 million children under the age of five die worldwide, almost 3 million in India alone. Over half of these deaths are caused by inadequate nutrition. At least two billion people suffer from vitamin and mineral deficiencies. If all the world's undernourished people were gathered together, the population of that "hungry continent" would exceed that of every continent except Asia.

AN ODYSSEY OF POVERTY

In his book *Earth Odyssey*, reporter Mark Hertsgaard eloquently describes the environment of the world's poorest. Having briefly shared life in 1991 with the Dinka tribe of sub-Saharan Africa, he writes: "The Dinka are a living reminder of the enormous environmental challenges human beings have faced on this planet since our emergence as a species untold thousands of years ago. At the end of the twentieth century, the Dinka are still living the way that virtually all of us *used* to live—as hunter-gatherers and small-scale agriculturalists on the edge of survival."[2]

The Dinka had been subsistence farmers in southern Sudan, one of the poorest places on earth. In the 1980s their already marginal existence was further eroded by a civil war that ravaged their homeland and forced them to flee their villages. Trekking two hundred miles into Ethiopia, they found shelter and survived for a time in a United Nations relief camp. But in May 1991 a violent coup in Ethiopia forced them to flee again, this time back into Sudan just ahead of their attackers. The Dinka's immediate plight was compounded by the chronic drought conditions that have plagued Africa for centuries. In this 1991 episode many of their numbers, especially the children, died of starvation, dehydration, and disease.

Hertsgaard tells us that

> the Dinka do not have the luxury of worrying about the environmental dangers of the twenty-first century, even though they are likely to suffer disproportionately from them: they have enough problems simply surviving from one day to the next. And the environment is no abstraction to them, as it is to so many people in the United States, Europe, and the rest of the wealthy, industrialized world. The Dinka experience the natural world directly, unmediated by electricity, running water, refrigeration, antibiotics, motor vehicles, and other modern technological marvels. Wildlife is the leopard that attacks their cattle or children, not something seen in books or at the zoo. And weather is no mere irritant to be neutralized with raincoats or central heating; it is an omnipotent unpredictable force whose whims determine whether there is enough food to eat.

Sub-Saharan Africa is the world's basket case of poverty, sometimes described as "the hopeless continent."[3] And the new millennium has not brought new hope to this region, only more despair. It is still the only place where hunger continues to increase in both the number and percentage of the population, reaching 180 million and 80 percent in 1990. Almost half the population lives on less than $1 a day. Child mortality before age five is the world's highest, and overall life expectancy for males, 44.8 years, is the world's lowest.[4] The "healthy life expectancy" of a baby born in Sierra Leone (in 1999) is only 25.9 years.[5] The school enrollment rate actually decreased during the 1980s in half the region's countries. Malnutrition is not declining, and one-third of children suffer from stunted growth. On average, the number of children per mother has barely declined in forty years and is still more than six, the highest of all the world's regions.[6]

How does it happen that the extreme poverty of sub-Saharan Africa stubbornly persists in an ever more affluent world? To what extent do environmental factors contribute to such poverty? And how does the poverty itself impact on the environment?

Nature has dealt an unkind hand to sub-Saharan Africa. The heat is intense and debilitating. The soils are typically poor and difficult to farm sustainably. The rainy seasons can be extremely variable, with recurrent floods in some places (e.g., Mozambique) and persistent drought in others (e.g., Ethiopia). The climate encourages insect-borne diseases such as malaria and dengue fever. Although most organized groups elsewhere in the world were historically able to cope with environmental hardships (the early Scandinavians, for example, adapted well to their long and cold winters), in Africa the environmental difficulties have been so severe that survival rather than development has remained life's main goal for many groups.

Yet nature's extremes, formidable as they are, do not alone explain the legacy of poverty and famine that still corrodes the environment of millions of Africans. Just as important are the centuries of slave trade and European colonialism (the latter ending only a generation ago), which sapped the land of its people and undermined its communities, institutions, and values and left an almost total vacuum of indigenous leadership and democratic tradition. While in recent times droughts and crop failures certainly have contributed to the region's chronic famines, civil strife is the source of many human disasters, the victims of which are mostly innocent civilians rather than combatants. The callous policies of many nondemocratic sub-Saharan regimes have also contributed to the environmental deterioration and social breakdowns, including unemployment and inequitable food distribution, that cause famines. All these factors have contributed to the region's enduring legacy of poverty.

Under such conditions, it is hardly surprising that environmental concerns considered important to many in the affluent nations, such as global warming and ozone depletion, are far off the radar screens of people living in the world's poorest places. If you happen to belong to the Dinka tribe, you probably have concerns of a much more immediate sort—for example, fear that your children may not survive even the next few weeks because they have been deprived of food, shelter, or medicine owing to bad weather or a new round of political repression. Despite international environmental festivities such as Earth Day and the many United Nations conferences aimed at impressing third-world countries with the importance of the North's environmental concerns, a genuine interest in these high-profile issues has not arisen in the developing world. Most of these countries are still in the first phase of their development, struggling to overcome the immediate challenges of survival. Although their peoples must depend on use of trees, soils, and water for survival, they have few incentives for conserving these resources because they neither own them nor benefit from their conservation. Under such conditions, people are not likely to show an interest in the environmental issues of the affluent until they themselves begin to taste the fruits of affluence.

THE SECOND PHASE OF DEVELOPMENT

In contrast to the prevailing situation in parts of sub-Saharan Africa, a number of developing countries elsewhere have passed beyond the survival barrier into the second phase of development—that of building a decent standard of living for their citizens through industrialization and modernization. On the Asian continent China and India are the largest and most visible of the "phase two" countries; Brazil is a good example in Latin America. A visitor to China will experience an environmental situation typically very different and less extreme than that of sub-Saharan Africa, yet one that is rife with environmental problems and equally revealing of the connection between poverty and the environment.

The economy of China is developing with breathtaking speed, and the same can be said of China's environmental landscape. Only two decades after the end of the economically and socially disastrous "Great Leap Forward" program imposed by Mao Zedong, China's major cities, such as Beijing, Shanghai, and Chongqing, have undergone amazing transformations, joining the ranks of the world's largest and most advanced metropolises. High-rise apartments, commercial centers, and industrial complexes proliferate endlessly; urban parks and green belts abound; and the automobile and freeway have become a fixture of the new cities. Among the urban populace, the growing business and professional classes are stylish, urbane, and consumerist, indistinguishable in many ways from their counterparts in London, New York, or Milan.

But China's urban environmental changes have also had a serious downside. Along with the proliferation of upscale buildings and shops, miserable shanty towns are rapidly appearing, housing the rural poor who flock to the big cities to try to improve their economic condition. But the most glaring environmental problem is the extremely high levels of air pollution from coal burning found in China's major cities. Visibility in some cities approaches zero during bouts of the most intense air pollution. Visitors to the capital city, Beijing, often develop bronchial inflammations after only a few days, especially if their visits come in late autumn or winter. Chinese citizens argue cynically about which city has the most polluted air, Beijing, or Chongqing in the south, or Benxi in the north. The air in these cities often contains levels of sulfur dioxide and respirable particles reaching ten times the maximum safe levels recommended by the World Health Organization—a truly unhealthy situation that can persist for days or weeks at a time. Compare this with the situation in Los Angeles, once one of America's most air-polluted cities, where the sulfur dioxide concentrations now remain well below the WHO and U.S. safe levels.[7]

Regardless of which city captures the dubious distinction of being China's most polluted, the causes of pollution are similar in all of them—rapid industrialization, skyrocketing electricity use, and almost total dependence on coal for electricity generation. Beginning in the 1980s, the growth in China's electricity use has been among the world's fastest, doubling approximately each decade, which reflects Chinese citizens' increasing ability to afford the benefits of adequate lighting and modern electric appliances. It is no wonder that coal is the major fuel for electricity production, since coal is China's most abundant energy resource and coal production already exceeds that of the United States.

Historically, coal has been the world's dirtiest fuel, and coal burning the world's leading source of air pollution. But this connection is no longer inevitable. Today, burning coal for electricity generation need not produce high levels of air pollution if state-of-the-art technologies are used for cleaning ("scrubbing") the exhaust stacks of the generating plants, a practice common (and legally required for new plants) in the United States and many other industrial countries. The problem is that China has rarely employed these advanced technologies, because they are so expensive to install and operate. For China at its present stage of development, achieving cleaner air (or other environmental benefits) has generally been of lower priority in allocating scarce financial resources than raising people's living standards by, for example, subsidizing traditional coal use to provide more and cheaper electricity.

In China, high levels of environmental pollution are found not only in the cities but also in many rural areas. Unlike the case in Africa, a great deal of industrial activity takes place in rural China. Thousands upon thousands of factories, from garage-sized plants to large industrial complexes, employ millions of skilled and unskilled workers, including our bicycle factory welder. The pollution sometimes takes the form of river contamination so severe that the waters become sickeningly unfit for consumption, yet such water is often used for drinking with only minimal if any purification. Rural water pollution in China is probably even less tractable than urban air pollution. The rural population not only is generally poor and uneducated, with little understanding of the health risks to which people may be exposed, but also is geographically very scattered and lacks influence with the environmental authorities. Even more unfortunate is that the rural working poor tend to accept their polluted environment as a symbol of, and a small price to pay for, the benefits of those millions of factory jobs.

There is growing evidence that this situation is changing, however, as both the Chinese economy and the Chinese people's environmental consciousness continue their fast-paced growth. Air-pollution control regulations are being enacted, and enforcement is being taken more seriously. In Beijing, clean-coal technologies are being installed and millions of tons of industrial coal are being replaced by natural gas. And now, galvanized by China's being awarded the 2008 Olympic games, the government is making earnest commitments to accelerate its clean-air programs. Given China's size and global importance, its environmental awareness, following on the heels of its increasing affluence, is a major reason for optimism about the world's environmental future.

DEVELOPMENT AS FREEDOM

Whether in sub-Saharan Africa, China, or elsewhere, chronic poverty deprives huge numbers of people of the incentives and economic means to care for and nourish their natural environment. Yet being poor is only one element of people's blighted relationship to the environment. According to economist and Nobel laureate Amartya Sen, poverty needs to be understood in broader terms than only the lack of monetary income. Sen argues that poverty should be characterized fundamentally in terms of the deprivation of basic freedoms, rather than merely low incomes.[8] In his view, development not only has the economic dimension with which it is usually understood but, more important, requires the removal of the "unfreedoms" endured by most people in underdeveloped countries. Besides poverty, these unfreedoms include deprivation of health care, lack of sanitation, exclusion from education (especially of women), exclusion from market activities, and above all, tyrannical regimes associated with systematic deprivation of political liberty and basic civil rights.

Development, in Sen's view, must include the freedom of democratic choices as well as the freedom of economic choices. Without such freedoms, people lack the opportunity for education, public debate, and discussion, which make possible rational choices about quality-of-life issues, including the environment, as well as rational choices about their own families or their government. It follows that environmental improvement requires not only a measure of economic power for individuals but also the broader freedoms of individuals to set priorities for themselves, their families, and their society. Such freedoms also nourish the development of social values and environmental ethics going beyond the bounds of government regulations and market rules. These values and ethics are essential for developing a healthy and sustainable environment.

I have argued above that countries and people in the earliest stage of development tend to have little interest in environmental issues as typically understood in the industrial countries, such as acid rain or global warming. In the subsistence phase, sheer survival amidst historically hostile environments has usually been the main challenge of their lives. This is not to say that poor societies do not have respect for their own environment—Native American nations, for example, generally have a profoundly spiritual relationship with their natural environment (though it has often been abused by outside forces). I have also noted that countries and people in the second phase of development, such as China, are quite aware of the collateral environmental deterioration occurring along with their industrialization and modernization. Yet in countries such as China, not only is domestic investment capital scarce, but also social priorities, including environmental quality, are set mostly by government rather than popular choice. Investments aimed at cleaning the environment typically have not reached the top of the government's priority scale, because other social investments (e.g., in energy production, housing, education, and industrial production for consumption and exports) have been seen as providing far greater benefits. As mentioned above, this situation is changing as China's economy rapidly grows.

ENVIRONMENTAL SEA CHANGE TO AFFLUENCE

A central thesis of this book is that the transition from the second phase to the third phase of a nation's development normally brings with it both a sea change in environmental consciousness and the political and economic means to care for and sustain a sound environment. In the Introduction, I traced briefly the environmental history of the United States and showed how this change took place as the country gained affluence following World War II. And it has just been mentioned that the same change is now happening in China—though decades will pass before the environmental improvements reach Western proportions. Of course I cannot assert categorically that people everywhere will automatically become protective of their environment as they become affluent, for that would stress my crystal ball beyond its capacity. Short of predicting the future, however, I cite the historical fact that a fundamental behavioral change toward environmental consciousness did take place in Western societies and Japan in the late twentieth century and is now beginning in China. I see no reason why we should not expect this to happen worldwide in the future.

In any case there is no mystery about the traditional connection between affluence and the environment. People of means have always sought to live amidst beautiful surroundings, regardless of the squalor that may have been nearby. And for most of history, it was relatively easy for the rich to isolate themselves from the environments of the poor by using fences, rivers, and other trappings of physical separation. Those eighteenth-century country estates of England were indeed magnificent examples of environmental isolation. But with the coming of industrialization, the rich had no possibility to fence themselves off from the flow of polluted air. Blackened with the coal smoke from the factories of London and Birmingham, that foul air was destined to be inhaled by rich and poor alike. One may surmise that the current concept of the environment as a collective resource, shared by all and the responsibility of all, was born at least partly out of that experience.

In this book I journey to those worlds apart—the environment of the poor and the environment of the rich. The journey allows us to look at the major environmental issues from both perspectives and provides evidence to support the argument that the most critical transition in the development of a sustainable future environment is the transition from poverty to affluence. This transition will obviously require at least several generations. Less obvious but no less important is the challenge to the global community to develop short-term environmental priorities that enhance the probability of long-term success as poverty is gradually reduced.

SIX BILLION AND COUNTING

Sometimes it seems that the world is just too full of people. Who has not fretted about overpopulation when pushing through teeming masses in a crowded third-world city? Or when trapped in a rush-hour sea of automobiles spewing exhaust gases from their powerful engines yet barely moving?

The specter of overpopulation has been a central theme of environmental pessimism for decades. Yet it is not only a recent concern; people have worried about overpopulation for centuries and have often speculated about how many people the earth can actually sustain. In a recent scholarly analysis, biologist Joel Cohen reviewed estimates of the earth's carrying capacity that range all the way from one billion people on the low side (which the earth surpassed years ago) to one thousand billion on the high side. (The present global population is six billion.) Cohen rejects the notion that this question can have a unique answer because a variety of evolving technical, sociopolitical, and economic factors, including lifestyle choices, will determine the bounds of the earth's population in any period.[1] The more important question may not be how many people *could* inhabit the earth but rather how many people are *likely* to inhabit the earth.

POPULATION GROWTH—GOOD OR BAD?

Demographic studies have become increasingly sophisticated, yet population growth remains one of the most controversial environmental issues. Opinions span the range from extreme optimism to extreme pessimism. The most optimistic view (detractors call it "cornucopian") holds that population growth is a *blessing* for humankind because each new person has the potential to become another Mozart, Rembrandt, or Einstein, uniquely capable of innovation and creativity. As far back as 1682, William Petty expressed the idea that "it is more likely that one ingenious curious man may rather be found among 4 million than 400 persons."[2] In this view, those who would halt population growth seriously undervalue the future contributions of people yet unborn.

In recent years the most persistent advocate of population growth's benefits was economist Julian L. Simon, who emphasized that the main contribution additional persons make to society is new knowledge, not only the kinds of knowledge provided by geniuses but also those provided by ordinary ingenious people.[3] And the more people, the better, according to Simon: more people create more knowledge and a demand for yet more knowledge. As Soichiro Honda, founder of the automobile company, put it, "Where 100 people think, there are 100 powers; if 1,000 people think, there are 1,000 powers."[4]

Of course, rich nations have a clear advantage over poor in putting those "1,000 powers" to work. In the rich nations, most people are given the tools of education so they can contribute to and make use of the growing stock of technological knowledge, which propels continuing increases in productivity and wealth. Lacking education and often freedom and opportunity, even the brightest individuals in poor nations are hindered from attaining and using the knowledge of which they are capable. Yet genius has a way of thriving in spite of severe handicaps, as witness the accomplishments of Beethoven, Helen Keller, and in our day, Stephen Hawking.

Simon correctly noted that people, especially experts, constantly underestimate the mind-boggling discoveries yet to be made. A stunning example of expert misjudgment is the 1943 remark attributed to Thomas J. Watson, then chairperson of IBM Corporation: "I think there is a world market for about five computers."[5] That might indeed have been the computer's destiny had not individuals been born into the world who invented the transistor and the integrated circuit, which allowed computers to become smaller, faster, and cheaper.

THE MALTHUSIANS

At the pessimistic extreme of the population debate is the notion that population growth is a terrible scourge upon humankind. This belief holds that global population will continue to grow until it is unsustainable, eventually crashing with disastrous consequences, including resource exhaustion and widespread famine and disease. The doomsday view of population growth originated in large part from the early pronouncements of the nineteenth-century English cleric Thomas Malthus, who believed that if people's natural procreational tendencies went unchecked, they would multiply to the point where the earth's food resources could no longer sustain them.[6] Malthus was convinced that populations tend to increase at a geometric rate (1, 2, 4, 8, 16, 32, 64, etc.), whereas available food resources grow only at a linear rate (1, 2, 3, 4, 5, 6, 7, etc.). He foresaw mass starvation followed by a horrendous population collapse when demand for food inevitably outstripped food supply. Less imaginative but no less pessimistic was the Italian economist Giammaria Ortes, who believed that the human population would grow until people over the entire earth's surface would be "crammed together like dried herrings in a barrel."[7]

Malthus and Ortes were both terribly wrong. But we shouldn't be too harsh in judging the doomsday prophets of those days, because they had no way of knowing that the social and economic forces that had always propelled humans to have large families were about to change radically. They could not foresee that medical science and sanitation, mainly through the control of infectious diseases, would soon free families from the timeless burden of bearing large numbers of children so that a few might survive to support them in old age. They also could not foresee the immense contributions that the technological revolution would soon make in increasing the efficiency of production in every area, including food. In fact, constantly advancing agricultural technologies allowed food resources (and the overall economy as well) to keep well ahead of population growth, increasing at a much faster rate than Malthus and others of his day believed possible. (In his later years Malthus changed his mind and withdrew his earlier doomsday prediction, yet it remains solidly associated with his name.)

In recent times the pessimistic view of population growth has been championed by biologist Paul Ehrlich, in a series of works beginning with his 1968 book *The Population Bomb*, in which he predicted that global overpopulation would cause massive famines as early as the 1970s.[8] Fortunately, famines of such magnitude never occurred.

Yet in 1968, fear about global overpopulation was not entirely without basis. Demographic studies then available showed that during the twentieth century population growth had in fact been extremely rapid by historical measures. From the year 1000 it took about five hundred years for the world's population to double; from the year 1500 it took about three hundred years to double again; from 1800 the next doubling took just under one hundred thirty years; and from about 1930 the next doubling took only about forty years.[9] If one simply extended the 1930–1970 population-growth trend line into the future, a very bleak picture would indeed emerge, with world population reaching fourteen billion by 2050 and thirty-two billion by 2100! If today we still considered such a future likely, we would certainly have something to worry about!

THE END OF POPULATION GROWTH

We can now be confident that such extreme population growth is not going to happen. Much more likely is that global population growth will slow and then cease altogether as the world moves from poverty toward affluence. The beginnings of the transition to a stable population are already quite in evidence. Global population growth has actually been slowing over the last two decades. Global population reached 5 billion in 1987 and passed the 6 billion mark in October 1999. The current growth rate is about 1.3 percent per year, which translates into a net global addition of 77 million people annually. On the basis of continually monitored demographic data, the United Nations now concludes that the growth rate will continue to decrease, and in consequence the UN's population projections have been steadily revised downward. In its 2000 revision, the UN projects a population of 9.3 billion for the year 2050 (middle-case projection), significantly lower than the 10 billion projected only four years earlier and nowhere near the 14 billion figure quoted above from the extrapolation of earlier trends.[10] The United Nations further projects that, with growth tapering off, the world's population will be almost static by 2100.[11]

Of course, one needs to exercise caution in interpreting the United Nations' statistical data. Extrapolating trends from the recent past into the future is always risky, as new trends can develop at any time if underlying conditions change. We saw above how extrapolation of the 1930–1970 population trend data produced an extremely high estimate of future population and kindled neo-Malthusian fears of global catastrophe. Thus one cannot absolutely rule out the possibility that the current worldwide downward trend in fertility will reverse and that population growth will start up again, even in the rich countries. But this turn of events is highly unlikely, as it is difficult to imagine a scenario in which the forces now driving population downward—improvements in income and health, educational and employment opportunities for women, birth control and family planning—are likely to reverse in the near future.

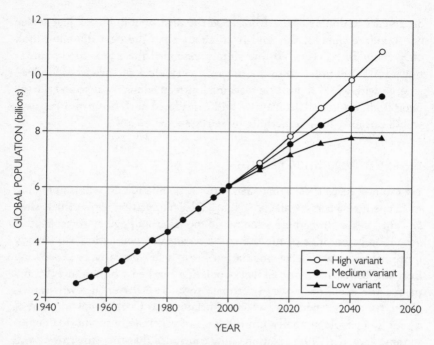

Figure 3. United Nations global population projections (1940–2060). Data are from *United Nations, World Population Prospects: The 2000 Revision* (New York: UN Population Division, Department of Economic and Social Affairs, 2000).

Throughout the world as people begin to live better (and longer), they are producing smaller families. Both current and historical data confirm this trend. Demographic data show that *high* fertility rates correlate with poverty and *low* fertility rates correlate with affluence. The current situation is illustrated by Figure 4, in which 1996 total fertility rates from a number of countries are plotted against their per-capita GNP (gross national product).[12] No country in the world today with a per-capita annual income over $5,000 (1994 dollars) has a total fertility rate significantly above the minimum replacement level (approximately 2.1 children per woman). This is a truly remarkable development, since an income of $5,000 is hardly what most of us would consider affluence, and even at that modest income level people are evidently choosing to have fewer children. Even in Sri Lanka, whose per-capita income is under $1,000, the fertility rate is only at replacement level. Surprisingly, sixty-one countries in the world are

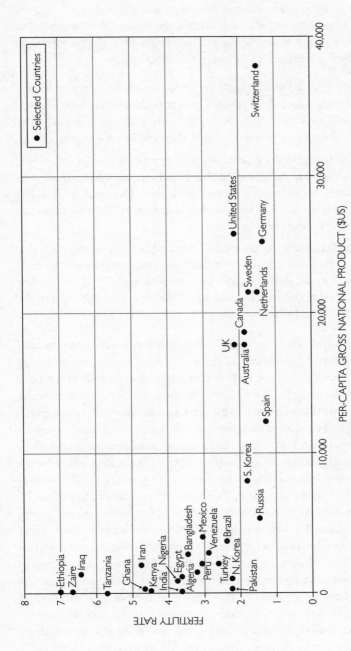

Figure 4.   Relationship between fertility rate and gross national product for selected countries (1996). No country with a per-capita annual income of $5,000 or higher has a fertility rate significantly above the minimum replacement level (2.1). Data obtained from the U.S. Department of Commerce, *U.S. Statistical Tables* (Washington, DC, 1996).

already at replacement level or below. In fact, if these countries do not soon increase their fertility rates, many will begin to *lose* population![13]

Although per-capita income is probably the best overall indicator of country fertility trends, a number of related factors also affect the kinds of decisions individual families make about their family size:

- INCIDENCE OF INFANT MORTALITY. As health-care improves and infant mortality decreases, parents gain confidence that most of their children will survive to adulthood. Thus they feel less compelled to bear more children as "insurance" that they will be cared for in old age.

- STATUS OF WOMEN. As women in poor countries gain freedom and education, they are increasingly able to achieve social status as individuals, often participating in the workforce. Hence they feel less desire to seek the kind of social status that historically was bestowed on women with large families.

- INDUSTRIALIZATION. As countries industrialize, the economic value of children changes. On the farm, children are an economic asset as helpers, but in an urban industrial setting, especially if child-labor restrictions exist, children become an economic liability because of the costs of raising and educating them.[14]

There is little doubt about the direct relationship between affluence and population stability. Nonetheless a "chicken or egg" argument persists as to which must come first. One view is that population growth in the poor countries must be brought down *before* they can develop their economies and move toward affluence. Some who support this view believe that coercive policies may be required to achieve a sufficient rate of fertility reduction.[15] Yet only one country, China, has actually mandated population control by law. During the 1970s several proscriptive norms aimed at reducing family size were instituted, and in 1979 a strict new policy sought to limit each household to only one child. The 1979 policy was implemented by levying fines according to how many children beyond the national guideline a couple bears. Though forced abortion or sterilization have never been included in China's official policies, allegations of coercion continue to be heard as local officials strive to meet population targets. According to one study, the one-child policy lost effectiveness after its first few years, as people resisted the policy and the government chose not to attempt strict enforcement.[16] In fact a dramatic decrease in the fertility rate was experienced in China during the decade *before* the one-child policy was instituted, falling from 5.8 (offspring per woman) in 1970 to about 2 in 1998.

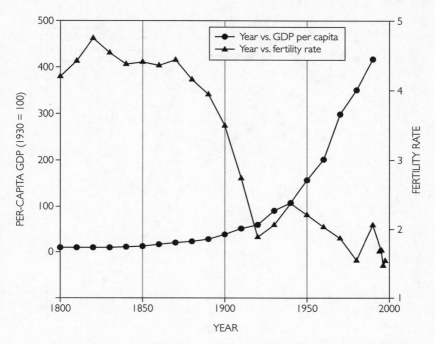

Figure 5.   Historical relationship between fertility and per-capita income in Sweden (1800–2000). The Swedish fertility rate dropped dramatically in the late nineteenth century as incomes (per-capita gross domestic product [GDP]) increased. The GDP data were provided by Professor Olle Krantz, Umeå University, Umeå, Sweden. Data on fertility are from Central Statistical Bureau of Sweden, *Historical Population* (Stockholm: Statistiska Centralbyran, 1997).

The fertility rate of China's urban population has actually now dropped below replacement level, reflecting the growing urban affluence and at least some of the preference factors cited above.

The experience of many countries, including China, suggests that harsh population-control measures are unnecessary for achieving population stability. Economic growth, urbanization, industrialization, women's liberation, health care—all these factors go hand in hand, each increase in the living standard stimulating families to have fewer children and the trend to smaller families itself lowering the economic barriers to affluence.

It is not unrealistically optimistic to point to the fact that in most countries poverty is really diminishing (though not as fast as we would like) and, with few exceptions, birthrates are really falling. The global fertility rate now stands at 2.7, probably an all-time low, and is continuing to decline throughout the world. As already stated, the United Nations has

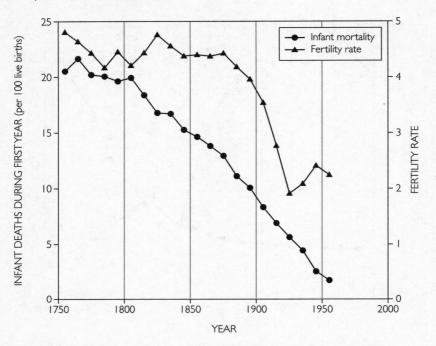

Figure 6.   Historical trends in infant mortality and fertility in Sweden (1750–1955). Data are from Central Statistical Bureau of Sweden, *Historical Population* (Stockholm: Statistiska Centralbyran, 1997).

with each passing year revised its future population estimates downward and now projects that most countries of western and northern Europe will experience actual population losses of 4 to 6 percent by 2050. And although the UN medium-case projection shows global population increasing until at least 2100, the low-case projection shows population peaking at about 7.5 billion in 2035 and actually decreasing thereafter.[17]

The inverse correlation between affluence and fertility is also apparent from historical demographic records, which in some countries extend back over two centuries. In Sweden, for example, the fertility rate steadily decreased after 1880, as income level rose and infant mortality fell. The fertility rate now stands at 1.6, well below the replacement value, 2.1. The growth rate of Sweden's population has been dropping since 1970, with the latest projections indicating that the population will top out around the year 2030 and slowly decrease thereafter.[18] The same dynamic is operating in other European countries. According to current estimates, twenty-eight

European countries will have a *lower* population in 2050 than today, while only ten countries will have a larger population.

A somber footnote to this discussion is the fact that the worsening AIDS epidemic has significantly lowered life expectancy in forty-five countries, thirty-five of which are in sub-Saharan Africa. (The worst hit are Botswana, South Africa, Swaziland, and Zimbabwe.) According to the United Nations, the combined population of these thirty-five African countries is projected to be 10 percent less by 2015 than it would have been without AIDS.[19] Nonetheless, by 2050 the populations of the most affected countries are expected to be larger than they are today. When population eventually stabilizes in these countries, as it no doubt will, the cause will surely not be famine and disease but rather health and affluence.

My bottom line is that population growth per se should no longer be looked upon as a serious long-term global problem, environmental or otherwise. The real problem is poverty. In the affluent countries, most people have already taken care of the historic population concerns by their spontaneous and independent decisions to have smaller families. People in the poor countries are already beginning to follow suit and will undoubtedly continue to do so as their conditions improve and their political and economic choices widen. So even though J. Simon may have been correct in pointing out that large populations could bestow some benefits on the planet, obviously people are deciding that smaller populations are more to their liking.

The global community, and especially the most affluent countries, has a profound moral obligation, through individual and collective actions involving both public and private institutions, to assist in bringing economic opportunities, education, and freedom of choices to people in the developing world. Population stabilization will be one reward. Benefits to the global environment, among many other rewards, will make this effort worthwhile.

3

CAN THE EARTH FEED EVERYONE?

Ask anyone who worries about the continuing destruction of tropical rain forests to identify the enemy, and the response is likely to include a sharp reference to farmers in places like Amazonia and sub-Saharan Africa who keep whittling away at the edges of the forests to create more agricultural land for themselves. Such an accusation would not be far off the mark, as small-scale agriculture is responsible for at least 60 percent of tropical forest depletion.[1] Why do people continue to commit such egregious offenses against precious forest ecosystems? The answer is poverty.

To poor farming households in many developing countries, keeping food on the table each day is life's primary challenge. Many subsistence farmers move to the forest's edge to escape poverty elsewhere, and they deforest to provide income and food for their families. Although these farmers may value the rain forest, they do not consider it valuable enough to be conserved for posterity at all costs.[2] From their perspective, and not unreasonably so, short-term survival has the highest priority, certainly a higher priority than long-term sustainability. So deforestation continues even today.

This apparent disregard for conserving threatened natural resources is a classic example of why poverty is so critical an environmental problem. Although these poor farmers benefit from degrading the forest, they do not pay, and in many cases are not even aware of, the costs that society may incur from their destruction of rain forests. The same can be said about poverty-related food-resource degradations such as salination, soil erosion, and overexploitation of fragile lands, especially those with low rainfall. Such local resource degradations are in many cases difficult, and almost always expensive, to reverse.[3]

These examples of cropland degradation by the poor are just the sort of scenario that the English cleric Thomas Malthus pictured in 1798, when he predicted that ever increasing numbers of people, motivated by short-term survival instincts, would consume critical resources faster than they could be replaced, and would eventually succumb to mass hunger and starvation.[4]

This kind of Malthusian scenario is evidently also what biologist Paul Ehrlich had in mind when he opened his best-selling book *The Population Bomb* with these words: "The battle to feed all of humanity is over. In the 1970s the world will undergo famines—hundreds of millions of people are going to starve to death in spite of any crash programs embarked upon now."[5] Fortunately a catastrophe of that magnitude never occurred in the three decades since the book was published. Still, local famines have continued, and millions of people have died of starvation—though for reasons more complex than the Malthusians suggest. The battle to feed all of humanity was by no means over in 1968. Nor is it over today. It will go on as long as vestiges of hunger and malnutrition degrade the human condition.

This battle not only can be won; it is being won. Through advances in science and technology over the last half-century, world food production has increased more rapidly than population, and food supplies have become more reliable, as well. Remarkably, the number of people worldwide suffering from acute malnutrition has fallen by three-quarters since 1960.[6] This is the story of the agricultural miracle called the Green Revolution. Starting in the early 1960s, new varieties of wheat developed by Norman Borlaug and his colleagues were shipped to Pakistan and India.[7] In Pakistan, wheat yields rose from 4.6 million tons in 1965 to 8.4 million in 1970. In India, yields rose from 12.3 million tons to 20 million. And the yields continue to increase. In 1999 India harvested a record 73.5 million tons of wheat, up 11.5 percent from 1998. The Green Revolution has helped transform India—a country where one of every three children died before age three—into a self-sufficient agricultural economy that has rid itself of the scourge of famine. Since 1968 India's population has more than doubled, its wheat production has more than tripled, and its economy has grown ninefold.[8] In fact, world grain harvests have more than doubled since 1960, and per-capita food production has increased by almost 25 percent.[9]

Today, neo-Malthusian predictions of global famine are inappropriate because sufficient progress has already been made to justify optimism about the world's food future. Even with expectations of continued population growth for several decades, hunger and malnutrition can be brought under control and eventually eliminated worldwide. The coming biotechnology revolution could be a crucial factor in the road to a sustainable global population and environment. In the spirit of the Green Revolution, this goal is being pursued around the world by many groups bringing together the best that agricultural science and technology have to offer.[10] The affluent as well as the poor stand to gain enormously from a world free of hunger.

WHAT IS THE REAL FOOD PROBLEM?

It is helpful to think of the world food situation as having not one but two distinct types of challenges. First, the challenge of total food *production*. How can the world produce enough food to feed every man, woman, and child at an adequate nutritional level, and do so in environmentally and economically sustainable ways? Second, the challenge of *distribution*. Even if total world food production is (or becomes) adequate, how will enough food be available reliably in the least-developed countries so that the poorest of the poor never go hungry? Although these two issues are in fact highly connected, let's first look at them separately.

CAN THE WORLD PRODUCE ENOUGH FOOD FOR EVERYONE?

The short answer is yes, the world can produce enough food for everyone, and using no more cropland than is presently under cultivation. In fact, since the mid-1970s the world has been producing enough food to provide everyone with a minimally adequate diet.[11] This conclusion conflicts with the widespread notion that the world is running short on cropland and that farmers everywhere will be forced to cultivate more and more land area, with catastrophic environmental impacts. True, in their present circumstances the poorest farmers are sometimes forced to resort to cultivating unsuitable areas, such as erosion-prone hillsides, semiarid areas where soil degradation is rapid, and cleared tropical forests where crop yields can drop sharply after just a few years. But these environmentally unsound practices will become less necessary as investments in modern agricultural science and technology begin to bear fruit, making possible continuing increases in efficiency of food production in the developing countries. The world is not short of cropland—it is short of affluence. A more affluent world will need less cropland, not more, to provide enough food for everyone. And an extra bonus will be the return of surplus cropland to nature, creating forests, meadows, and parks.

But we need to ask: how much food is "enough"? According to the United Nations Food and Agricultural Organization (FAO), a daily diet of about 2,300 calories per person would provide adequate nutrition if everyone, especially in developing countries, had equal access to food.[12] FAO calls this nutrition level the "national average daily requirement." Adjusting for a moderate level of diet inequality, FAO raises the average daily requirement to about 3,000 calories per person.[13] The total global food requirement can be calculated by multiplying the number of calories per person times the global population. So, if all the world's expected 9.3 billion inhabitants[14] consumed food at the 3,000-calorie daily level, the total annual food requirement would be about 10^{16} calories (or, ten million billion calories).

For this amount of food to be produced from the area of cropland currently under cultivation (1.4 billion hectares, or 11 percent of the world's land), the annual crop yield would have to be, on average, about seven million calories per hectare of cultivated land. This is equivalent, for most crops, to a yield of about 1.8 tons per hectare. For comparison, the current yield of wheat in some arid regions of Africa is about 1 ton per hectare. It would take only twice that yield for the current cropland to provide a 3,000-calorie diet for the world's expected 9.3 billion inhabitants.

The industrial countries, of course, do considerably better than this even today. Wheat is produced currently at an average yield of 3 tons per hectare in North America and 6 tons per hectare in Europe. Maize is produced in the United States at an average yield of 8 tons per hectare, and prize-winning yields as high as 20 tons per hectare have been achieved.[15] Rice is produced in South Korea at an average yield of 6 tons per hectare.[16]

In the coming decades, application of new agricultural technologies and improved management of agricultural systems have the potential to raise agricultural productivity considerably in the developing countries. For example, recent production with available technology in the Brazilian savannah, whose acid soils were hitherto thought to be worthless for agriculture, has given yields of 6 tons per hectare when irrigated and 3 tons when rain fed.[17] In the industrial countries, smaller future gains are expected because the average productivity is already very high. (Note that there is still considerable room for the average productivity to creep up toward the maximum genetic potential.) The bottom line: if the world's average crop yield reaches only that of wheat production in the United States today (3 tons per hectare), the world's nine billion people could enjoy an average daily consumption of 3,000 calories and use less than two-thirds of the land area under cultivation today.[18]

Calories from cereal grains are, of course, not the whole story of food sufficiency. The diets of today's affluent consumers in North America and

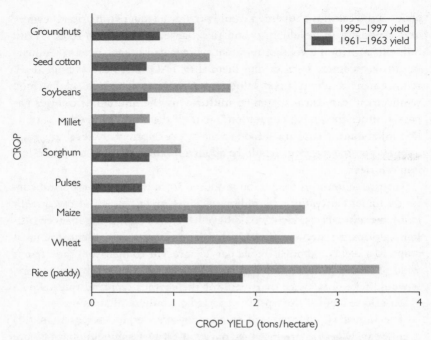

Figure 7.   Major crop yields in developing countries, showing production increases from 1961–1963 to 1995–1997. Data are from United Nations Food and Agricultural Organization, *Agriculture: Towards 2015/30*, technical interim report (Rome: FAO, April 2000).

Europe not only include cereals, vegetables, fruits, and milk products but derive as much as 30 percent of their calories from animal products, which supply needed proteins and micronutrients as well as calories. If diets in the developing countries improve to the point where people strive to emulate today's European and North American diets, world food demand could be considerably higher. The next global revolution in food production and consumption is sometimes referred to as the Livestock Revolution because meat and milk will increasingly substitute for grain in the human diet, requiring increasing amounts of cereal-based animal feeds.[19] Yet for the foreseeable future, it is reasonable to assume that the caloric increases needed to eliminate chronic hunger will come mostly from plants, especially cereals.

Projecting food requirements should take into account not only efficiency of food production but also efficiency of consumption. End-use losses from spoilage, processing, preparation, and plate waste are estimated to amount to as much as 70 percent of the food actually consumed.[20] Some of these losses can be expected to decrease with increasing affluence because of better shipping and processing technologies; nonetheless, affluent countries such as the United States, Belgium, Switzerland, and Italy continue to waste nearly 60 percent of their food. If such losses were cut in half, global food requirements would decline by over 7 percent.[21]

Overall, the world food situation looks bright because of the astonishing advances already made in agricultural productivity in both the developed and developing countries. Especially encouraging is the fact that the gap in average cereal yields between the developed and developing countries is beginning to narrow. Dr. Borlaug puts the productivity issue in perspective:

> I have calculated that if the United States attempted to produce the 1990 harvest of the 17 most important crops with the technology and yields that prevailed in 1940 it would have required an additional 188 million hectares of land of similar quality. This theoretically could have been achieved either by plowing up 73 percent of the nation's permanent pastures and rangelands, or by converting 61 percent of the forest and woodland area to cropland. In actuality, since many of these lands are of much lower productive potential than the land now under crops, it really would have been necessary to convert an even larger portion of the rangelands or forests and woodlands to crop production. Had this been done, imagine the additional havoc from wind and water erosion, the obliteration of forests, and extinction of wildlife habitats, and the enormous reduction of outdoor recreational opportunities.[22]

As encouraging as the progress has been, agricultural productivity gains have not yet produced a decisive victory in the battle against hunger and malnutrition, and a dark cloud hangs over one developing region, sub-Saharan Africa, which continues to lag behind the other regions, particularly East Asia, in grain productivity increases.[23] For that decisive victory to be achieved, the food productivity goals made possible by advanced agricultural technologies *must be met* during the coming decades and *must be sustained* afterwards. Without doubt, the real world presents many obstacles standing in the way of developing countries achieving and maintaining these gains. Some of these obstacles relate to the possible limitations of agricultural and ecological science, but others—perhaps even more challenging—are posed by lack of freedom and education, by war and civil conflict, and of course by the enormous barriers raised by poverty itself.

CAN THE WORLD DISTRIBUTE ITS FOOD SUPPLY TO EVERY PERSON?

As discussed above, enough food is produced globally today to feed everyone if it were evenly distributed. Unfortunately that is not the case, as evidenced by the continuing occurrence of malnutrition and hunger in developing countries and, to a much lesser degree, in some affluent countries. What causes such wide disparities in food availability in the midst of plenty?[24]

Thus far in this chapter we have looked at the food-production problem in global terms because that helps to clarify the larger issues of agricultural land use and the status of agricultural science and technology, besides helping to track progress in food productivity through time. But food availability is very much an issue localized to individual countries or regions. For one thing, access to food is largely determined by income level, which is locally determined. Insufficient family income is the proximate cause of much of the world's hunger and malnutrition, especially when famines strike. About 1.3 billion people live in households with daily earnings less than one dollar per person, while another 2 billion people are only marginally better off.[25] In sub-Saharan Africa three-quarters of the population live in such poverty. These poor households cannot afford to buy enough food even when the markets are well stocked. It is paradoxical that this situation exists even though over the last thirty years, real food prices have fallen by over 50 percent.[26]

Second, low income is only part of the problem in the developing world. At a more fundamental level, political powerlessness is a root cause of the hunger that still plagues citizens of particular developing countries. In many developing countries, especially in sub-Saharan Africa, the political structure renders poor people powerless, and it is nearly impossible for impoverished people, no matter how numerous, to mount the political strength required to force their governments to adopt policies that fight hunger effectively and promote a more equitable income distribution. Yet there are contrary examples: in Brazil, Zimbabwe, and the Indian state of Kerala, popular movements have pressed governments to end hunger. In South Korea the government has enacted public policies that fostered economic growth accompanied by decreasing income inequality.[27]

Third, in the poorest countries, critical natural resources, including fresh water, land, forests, and fisheries, are being used at or beyond capacity. In the competition for resources, poor and hungry people, lacking economic and political clout, become even more marginalized. Especially in countries where landholdings are inequitably distributed, poor families try to eke out a livelihood by moving onto fragile land and often into over-crowded cities.[28]

Yet probably the most pervasive and stubbornly intransigent cause of hunger is localized military and civil conflict. A large number of developing countries have suffered from internal and external conflicts that plunder food supplies, interrupt food cycles, and destroy seeds and livestock. These conflicts may have produced as many deaths from starvation, especially among children, as from combat. The list includes, among others, Afghanistan, Burma, Mozambique, Rwanda, Somalia, Sri Lanka, Sudan, and certain Balkan countries. Even in the absence of actual conflict, the developing countries annually spend more than $100 billion on military expenditures. One-quarter of this amount could provide their citizens with primary health care, family planning services, and adult education.

FAMINES

Famines are an acute and severe form of the chronic food insufficiency that plagues a number of developing countries. A famine is more than a localized food shortage—it is an environmental disaster, a total disruption of the systems and institutions that produce and distribute food. Recurring famines are responsible for loss of countless lives each year in developing countries, especially in sub-Saharan Africa.

Droughts or floods are sometimes the immediate triggers of famine, but the root causes are invariably deeper. Mostly famines are associated with undemocratic and oppressive regimes whose policies, or lack of policies, create vulnerabilities among the poorest citizens that weaken their resistence to the onslaught of famine conditions. And more often than not, those regimes are also involved in armed conflict, internal or external, which greatly exacerbates the marginal conditions in which the poorest families live. Without resolving the underlying causes of their political conflicts, the African countries will enjoy little success in reducing the incidence of famine.

But apart from civil and military conflict, poverty remains famine's most basic cause. And the manifestations of poverty are similar throughout the sub-Saharan region: primitive technological status, lack of educational and employment opportunities, oppression of women, and dreadful environmental conditions, especially regarding water, sanitation, and health care. People living in such conditions are very poorly equipped to resist the shocks accompanying occasional extreme weather or harvest conditions, and it is no wonder that famines spread like contagious diseases in such circumstances.

Famine is the localized case in extremis of the general worldwide food security problem. Enough food is produced today to feed everyone, and in coming decades there will probably be more than enough to go around. But achieving global food security, as we have seen, involves more than achieving global production goals. It also implies achieving and sustaining sufficient production locally in the most vulnerable places and also maintaining a distribution system that does not fail the most vulnerable families when difficult times arise. To achieve an adequate famine prevention system requires both comprehensive agricultural development, which is achievable, and far-reaching political progress, which is more problematic. The authors of the book *Famine in Africa* summarize the situation succinctly:

> There is no excuse for the continued occurrence of famine in today's world. Famine represents a failure of politics and action at every step, from the local to the international community. Public intervention based on partnerships between communities and government agencies can and does effectively overcome famine. The citizens of famine-prone countries have a right to expect measurable progress toward such a goal. Having enough to eat is not just an abstract human right, it is the basis of all functioning of society and hence must be the foundation for sustainable development.[29]

THE REAL CHALLENGE: SUSTAINABLE INTENSIFICATION

For centuries, food production increases in agriculture came about mostly from increasing the *acreage* of farmed land. Today, most of the world's available land is under cultivation, so this option is not expected to contribute much in the future. The Green Revolution's remarkable increases in food production in developing countries came about primarily from agricultural *intensification*—coaxing more and more food to grow on each plot of cropland rather than adding more acreage of cropland.

What were the factors that actually produced the Green Revolution's agricultural intensification and doubling of world grain production? These stand out:

· Low-yield grain varieties were replaced with high-yield varieties bred for resistance to pests and diseases.

· Industrially produced chemical fertilizers and pesticides were increasingly applied.

· Irrigation was much more widely used.

· New agricultural machinery and techniques were introduced.

The results of the intensification were spectacular: rising yields for important crops and increased numbers of crops that could be harvested each year.

Looking to the future, agriculture needs to find ways to continue the intensification started by the Green Revolution if crop production is to keep ahead of the next half-century's population growth and to sustain that intensity after global population stabilizes in the latter part of the century. Simply put, in the future more food must be grown from each unit of land, water, energy, and labor invested in agriculture.

Though most agricultural scientists agree on the nature of the challenge, they are deeply divided over the proposition that intensification can continue for another half-century. Gazing into a crystal ball brings out the optimist or pessimist in most people, and scientists are no exception.[30] The two "groups" ponder exactly the same body of agricultural information, and mostly they agree about the facts and uncertainties of the data. Where they differ is about the significance of what is being measured. For example, one group of economists and ecologists argues that data indicating gains in production of commodities such as cereals may give a falsely optimistic picture of the future, because the data may not include concomitant losses in the natural resource base on which that production depends. This omission can occur because some natural resources (e.g., ecosystem services) are not accounted for by standard economic measures.[31] It is their concern about the sustainability of the natural resource base that leads these scholars to their skeptical view about the sustainability of agricultural intensification.

Currently there are signs that intensification may in fact be slowing down. Although crop yields are generally still increasing, in some crops the *rate of increase* has shown signs of tapering off. For example, worldwide productivity (yields per hectare) of cereals harvests increased at a rate of only 1.5 percent per year from 1982 to 1994, compared with a 2.2 percent per year rise from 1967 to 1982.[32] Does this suggest a temporary trend or a long-term trend downward? If you are a pessimist, you worry that this trend may presage a total cessation of yield improvements; while optimists point to projections indicating that world food supply will continue to outpace population growth at least until 2020 and that per-capita availability of food will increase by about 7 percent by then.[33]

Thus it is that individual scientists can look so differently at some of the factors that may ultimately determine the world's food future. Another important area where differences appear is fertilizer use, which is widely projected to increase significantly to provide adequate soil nutrients for future intensification. On the pessimistic side, concerns are expressed about possible exacerbation of known biological and environmental impacts from fertilizer use, including nitrate contamination of water supplies and promotion of destructive insect populations. Some concerned scientists recommend that the current growth in fertilizer use be slowed, while others go further and urge that fertilizer use be drastically cut back. A more optimistic, technologically based view holds that fertilizer use can be safely increased by improving the ways fertilizers are applied, moving away from today's brute-force techniques to much more selective applications—for example, by applying fertilizers differentially to match crop demand and by timing fertilizer applications precisely to track the plants' nitrogen needs.

Strong differences also emerge about irrigation, a practice destined to increase because of its crucial role in intensification. There are serious concerns about irrigation's high cost and about chronic water shortages in developing countries. The optimistic view focuses on available technological improvements such as precision drip irrigation and high-efficiency sprinklers, which can markedly increase the efficiency of water use. But the skeptics question whether doubling of today's irrigation efficiency can actually be achieved in countries where it is most needed and also express concern over the degradation of irrigated agricultural land throughout the world by salinization, water logging, and alkalization, all of which reduce agricultural productivity. These problems can usually be mitigated but typically at high monetary and energy costs. Here optimists point to a variety of technological improvements that may reduce those costs.[34]

BIOTECHNOLOGY AND THE SECOND GREEN REVOLUTION

Biotechnology involves the use of molecular gene-splicing techniques to optimize living systems to provide better drugs, foods, and other products while reducing or eliminating undesirable features. In the industrial countries, the first two decades of the biotechnology revolution have brought forth a remarkable collection of new diagnostic tools, medicines, and medical therapies aimed at prevention and treatment of human diseases.[35] Judging by how well these medical products have fared in the commercial markets, one could say that the future of biotechnology looks very bright.

The context in which biotechnology developed in the affluent countries, however, is so different from that of the developing world that one can justifiably question the relevance of current biotechnology to the problems faced by the world's poor. Yet few earthly needs are more urgent than applying biotechnology's incredibly innovative science to the developing countries' struggles against poverty and hunger. The affluent world has an obligation to ensure that modern biotechnology does not bypass the poor farmers and consumers of the developing world.

The revolution of agricultural biotechnology—the Second Green Revolution—is well underway in the industrial countries. Biotechnology research is generating the knowledge that will make possible the production of plants with higher production yields, greater resistance to stresses, and lower requirements for inputs of environmentally toxic chemicals. In the United States, transgenic varieties and hybrids of cotton, maize, and potatoes containing genes that effectively control a number of serious pests are being introduced commercially.[36] Already in 1996, 1.7 million hectares were planted with transgenic crops worldwide. In 1998, this acreage had jumped to 28 million hectares, about 60 percent of which is in the United States, China, and Latin America.[37] Although no one expects gene technology to be the silver bullet that by itself can save the world from starvation, its potential for increasing the quantity and quality of crops grown in the third world is enormous.[38] This potential and the progress already achieved are reasons why I could confidently write, earlier in this chapter, that the battle against hunger is being won.

In the developing countries, applications of biotechnology research are being targeted to high-priority food-security problems, especially the production yields of grains, meat, and milk. In China a big jump in rice productivity may be just around the corner if current research in Hunan province succeeds in creating a super-high-yield hybrid that promises 15–20 percent increases in rice yields over existing hybrids.[39] Hybrid rice already accounts for half of China's rice acreage and yields an average of 6.8 tons per hectare compared with 5.2 tons for conventional rice, the increased output feeding an additional hundred million Chinese each year.

Rice is the most important staple crop also in Costa Rica, providing almost one-third of the daily caloric intake. Production costs have been increasing because of growing pesticide and fungicide use, yet yields have remained static. A biotechnology program aimed at increasing rice biodiversity features a strategy that includes the possible use of native wild-rice germ plasm, which may harbor useful agronomic traits for use in crop improvement.[40]

In Thailand the shrimp aquiculture industry saved over $500 million in 1996 through diagnostic DNA research that reduced chronic losses from shrimp viral pathogens.[41] Thailand also produces a high-quality aromatic rice that could be a contender in world markets if low yields caused by blast disease can be overcome. Research is underway to identify genes that would confer resistance to this disease.

In Hawaii, a cooperative project with Cornell University has developed transgenic papayas resistant to the ring-spot virus. As a result of this research, the papaya industry was recently rescued by introduction of a genetic "vaccine" that immunized papaya trees against the ring-spot virus, which was destroying the entire crop.[42] This research is making possible the reintroduction of papaya cultivation to small farms in areas where the crop had previously been decimated by this disease. Similar research on common beans is aimed at breeding resistance to the golden mosaic virus.[43]

Throughout the developing world, genes producing beta-carotene, a precursor of vitamin A, are being inserted into rice to produce a new variety of golden rice that could prevent millions of cases of blindness and death in children suffering from vitamin-A deficiency.[44] Generated by a Swiss research team, this rice is being distributed without charge to public rice-breeding institutions around the world, which will incorporate the new rice traits into local rice varieties for growing by local farmers.[45]

These examples show two things. First, that serious efforts are underway in developing countries to apply the industrial world's biotechnology knowledge to their own pressing agriculture problems. Second, that the scale of these efforts is still minuscule in comparison with the need and with the potential of biotechnology in the developing world. For this potential to be realized, those who are dedicated to a future sustainable world—governments, institutions, enterprises, individuals—should put their shoulders and their wallets behind this enterprise. The importance of support is underscored by a simple economic reality: third-world farmers live largely outside the market economy and will rarely be able to afford the products of biotechnology research, most of which will be marketed by transnational agribusinesses. If small-scale farmers in the poor countries have a right to share in the benefits of biotechnology, which they surely do, the affluent world is obliged to extend a helping hand.

THE ANTIBIOTECHNOLOGY MOVEMENT

Rather than extending a helping hand to biotechnology, however, some are extending a clenched fist. A strong antibiotechnology segment of the Green movement seeks to discredit and eliminate the development and use of biotechnology. The opposition to biotechnology is based on exaggeration of the risks of genetically modified organisms and denial of the benefits. In fact, the risks of biotechnology are very small and the potential benefits are enormous. Nor is there anything new about genetically modifying organisms. Almost all of our traditional foods are products of natural genetic mutations or genetic recombinations. For thousands of years— ever since human agriculture began—plants and animals have been genetically modified by selective breeding, giving us beef, wheat, corn, oats, potatoes, pumpkins, rice, sugar beets, and grapes, with no evidence of harm to either public health or the environment. Whatever risks there may have been in traditional selective breeding—and these were very small—the risks from adding specific genes via genetic engineering are even smaller since the products can be much more precisely controlled. In any event, since 1994 three hundred million North American consumers have been eating several dozen genetically modified foods including canola, corn, potato, papaya, soybean, squash, sugar beet, and tomato, grown on hundreds of million acres—with not a single documented problem.[46]

The genetic modifications (GMs) in these crops have provided a number of benefits to farmers and consumers. GM has, for example, given enhanced herbicide resistance, which decreases competition from weeds and allows fewer herbicides to be used, lowering costs and raising quality. GM has provided increased resistance to insects and diseases, which boosts crop productivity while also lowering costs. GM has been used to delay the ripening process of tomatoes, prolonging shelf life and facilitating harvesting and transport to markets. In the case of soy and vegetable oils, GM has reduced saturated fat content and, in one soy product, increased the desirable monounsaturated, fatty oleic acid from about 24 percent to over 80 percent. Many other advances are forthcoming, including enhanced flavor, texture, and nutritional value; reduced absorption of fats in frying; increased use of desirable enzymes in food processing and aging of cheeses; lowered calorie content of beets; and reduced allergenic components of foods such as peanuts.

Genetic modification has probably been more thoroughly scrutinized than any prior crop-breeding technology in the history of agriculture. For years the safety of genetically modified food products has been under constant examination by government and university scientists in many countries. Certainly some food products have inherent risks, for example, the risk of excessive toxic alkaloids in tomatoes or allergens in Brazil nuts. But these risks are the same whether the crop is produced by traditional or modern technologies. No specific risks or harm have been identified from the genetic modification process itself. Were there any inherent problems with GM technology, they would almost certainly have been revealed by now. But not one problem has been documented.[47]

The case of food allergies is interesting, because opponents of genetic modification claim that allergens are a serious risk of GM food. This claim is based on misinterpretation of research results showing that food properties, including allergens, can be transferred from one species to another. No commercial food products were involved in the research. Actually the relationship of GM food to allergens is quite the opposite: scientists' new ability to identify specific genes responsible for allergic reactions in particular foods can be used to *remove* those genes. In the future we can expect to see nonallergenic GM peanuts, dairy products, cereals, and seafood on grocery shelves.

In a widely publicized misinterpretation of preliminary laboratory research, an anti-GM advertisement stated: "Cornell University scientists discovered that genetically engineered [Bt] corn pollen killed 50 percent of Monarchs [butterflies] in their test."[48] In fact, the preliminary experiment referred to lacked controls, and the effect of GM pollen on Monarchs was subsequently found to be negligible under field conditions.[49] And the Monarchs appear to be doing very well, as measured by the numbers arriving in Mexican sanctuaries in spite of the fact that almost a third of the U.S. corn acreage is now planted with genetically engineered Bt corn.[50]

Most scientists knowledgeable about genetic engineering recognize the false assumptions underlying most antibiotechnology claims, and they are confident that the potential benefits far outweigh possible risks. A petition signed by over twenty-one hundred scientists worldwide, including Nobel laureates James Watson, codiscoverer of the DNA structure, and Norman Borlaug, father of the Green Revolution, begins: "We, the undersigned members of the scientific community, believe that recombinant DNA techniques constitute powerful and safe means for the modification of organisms and can contribute substantially in enhancing quality of life by improving agriculture, health care, and the environment."[51] In April 2000 a U.S. House of Representatives report concluded that there is no significant difference between plant varieties created by agricultural biotechnology and similar plants created by conventional crossbreeding.[52] And concurrently a U.S. National Academy of Sciences committee concluded, "The committee is not aware of any evidence that foods on the market are unsafe to eat as a result of genetic modification."[53]

Arguments against biotechnology are usually couched in terms of the so-called "precautionary principle." Of course, few would disagree with the notion that precaution is in order when one confronts an unknown or risky situation. People were understandably cautious about the risks of early automobiles, early airplanes, and even early electric lightbulbs. And since September 11, 2001, people have understandably become cautious about the risks of terrorism. But when opponents of genetically modified foods tell us, under the guise of earth-friendly advocacy, that we should say *no* to this or any technology that cannot absolutely be guaranteed to be without risk, they are perverting the precautionary principle into an instrument for creating fear of innovation.[54] Cliches such as "you can't be too careful" and "better safe than sorry" completely ignore the common-sense appreciation that no human activity can be totally risk free. If the same logic were applied to the risks associated with automobile travel, which, though small, are much larger than those of genetically modified foods, no one would ever ride in an automobile.[55]

Science cannot guarantee absolute certainty. But science can and does allow us to compare the risks of alternative human actions against their benefits. The alternative promoted by most opponents of genetically modified foods—an indefinite worldwide moratorium or outright ban—carries the risk of a world increasingly unable to meet the nutritional needs of all its human inhabitants. That risk far outweighs any possible benefits of such a ban and, on moral grounds, is unacceptably high.

It would be a pity if fear of genetically modified foods were to cause environmentally conscious citizens, genuinely abhorring the plight of the poor, to contribute unwittingly to denying the developing nations in Africa and Southeast Asia access to decades of research and discovery that could help them produce more and better food, in effect condemning millions of the world's children to continuing malnutrition, hunger, and disease.

Nobel laureate Norman Borlaug comments on the antibiotechnology movement:

> I now say that the world has the technology—either available or well-advanced in the research pipeline—to feed a population of ten billion people. The more pertinent question today is whether farmers and ranchers will be permitted to use this new technology. Extremists in the environmental movement from the rich nations seem to be doing everything they can to stop scientific progress in its tracks. Small, but

vociferous and highly effective and well-funded, anti-science and technology groups are slowing the application of new technology, whether it be developed from biotechnology or more conventional methods of agricultural science. I am particularly alarmed by those who seek to deny small-scale farmers of the Third World—and especially those in Sub-Saharan Africa—access to the improved seeds, fertilizers, and crop protection chemicals that have allowed the affluent nations the luxury of plentiful and inexpensive foodstuffs which, in turn, has accelerated their economic development. While the affluent nations can certainly afford to pay more for food produced by the so-called "organic" methods, the one billion chronically undernourished people of the low-income, food-deficit nations cannot.[56]

In the affluent countries, pharmaceutical and medical diagnostic applications of biotechnology have been enthusiastically received because the public understands and appreciates both their success and the potential for even more remarkable disease-conquering products. With the continuing accumulation of evidence for the safety and efficiency of biotechnology in agriculture and the continuing absence of evidence of harm to the public or the environment, most consumers in the affluent countries will increasingly welcome the growing array of genetically enhanced food products. But for billions of farmers and consumers in the developing countries, the Second Green Revolution could be much more than a welcome addition to their food menu—it could be the prime agent of a better life and the saver of hundreds of millions of lives.

4

FISH TALES

In an influential 1968 paper, biologist Garrett Hardin proposed a simple explanation of why we humans despoil our environment: because the environment is a "commons," belonging to everyone and therefore to no one.[1] Hardin asserted that overexploitation of freely accessible resources held in common is virtually inevitable, and he referred to the social cost of such overexploitation as "the tragedy of the commons."

Years ago, some friends and I hiked to a remote country lake where we caught a few fish for a delicious dinner. These fish had been owned by no one until they became ours, and of course we paid nothing for them. Yet our removing the fish from the lake actually did carry a social cost—the small reduction of the fish population for future fishing. At that time, the lake had very few visitors, and I would have judged that cost to be insignificant had I even thought about it. Eventually, however, that area became less remote and more accessible, hiking increased, and the lake became popular for sport fishing. In time, fish were being removed from the lake faster than the stock could replenish, and it wasn't long before "no fishing" signs appeared all around the lake. As of now, people are still denied the pleasure of fishing in that lake, because too few fish remain. Does this little fish tale illustrate an inevitable fate of the world's fisheries?

Not necessarily. Although few would deny that many of the world's great fisheries are in trouble, there is mounting evidence that they can be restored to their former health and that fishing can make a major and sustainable contribution to the world's future food supplies. Also, fish farming, or aquiculture, has great potential to further supplement marine fishing as a major food resource. But today, as a consequence of recent overfishing, the environmental problems related to marine fisheries remain serious. It's important, though, that these problems be kept in perspective. Recall that the air and water environments of many nations suffered serious degradation as they rapidly industrialized. In recent decades those nations, as we have seen, made great progress in restoring the quality of the environments within their borders. Because many fisheries are an international resource—a global commons—they have not, until recently, enjoyed the benefits of institutions strong enough to ensure their sustenance. Although the world has not yet decided with one voice how best to treat this patient, it is far too early to write an obituary for the world's great fisheries.

The oceans are among the world's greatest commons—they are owned by everyone and by no one.[2] Throughout history, ships and their intrepid sailors from many countries enjoyed "freedom of the seas." Fish, a natural and mobile resource in the open oceans, were always considered common property, to be taken freely. Those engaged in commercial fishing were concerned about maximizing their own catch, and few gave thought to the possibility of overexploiting the fish stocks. In any case, so long as the fish kept coming in, there was no reason for people to doubt that the stocks were inexhaustible—especially since humans had no reliable methods of gauging the sizes of various fish resources. Fishing thrived for centuries, and in the great fisheries off the New England coast vast catches of cod, haddock, and flounder provided livelihoods for thousands and food for millions. Today, fisheries provide direct employment to about two hundred million people globally, and account for 19 percent of the total human consumption of animal protein.[3] But just as in that little mountain lake, overfishing is threatening the sustainability of this precious resource.

OVERFISHING: WHY IS IT HAPPENING?

Before World War II, fishermen had enjoyed open access to most of the world's fisheries. As if to affirm the oceans' bounty, the total global catch of fish continued to grow year after year. And after the war, advances in navigational equipment, fishing gear, and harvesting techniques increased both the safety and productivity of fishing. But costs and consumer demand also steadily increased, which created economic pressures for increasing the harvests. From 1950 to 1990 the number of vessels in the world's fishing fleet more than doubled, and a serious overcapacity in the fleet developed. This created ever more intense competition over the fixed resource. In response to the increasing harvesting pressures, the global fish catch

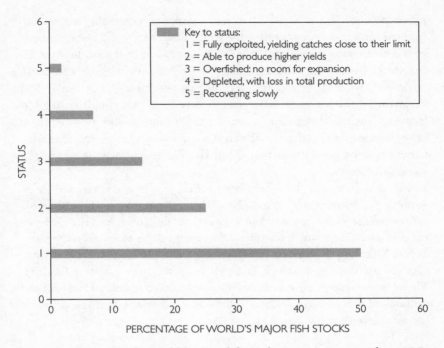

Figure 8.   Status of the world's major fish stocks (1990s). Data are from UN Food and Agricultural Organization, *Agriculture: Towards* 2015/30 (Rome: FAO, 2000), chapter 7.

more than quadrupled. Yet profits fell as costs continued to escalate. The United Nations Food and Agricultural Organization (FAO) estimated that in 1993 the world's fishing fleet catch was worth $72 billion but cost $92 billion to catch. Today the fishing fleet's capacity is more than twice as large as would be needed to fish the oceans sustainably. Under free-market conditions the overcapacity would diminish naturally as the least profitable enterprises decided to leave the business or went under financially. But free-market conditions have not existed for decades in fishing; instead, the ailing industry is being sustained by tens of billions of dollars of government subsidies, which only exacerbate the problem by providing incentives for further fleet expansion. And especially in the large international fisheries, the economically and culturally diverse competitors also contribute to overfishing because they have little tradition of cooperation in managing the resource that provides their livelihood.[4]

The result of all these factors is a natural resource that is under great pressure and is unsustainable under present conditions. In 1989 total world production of fish and shellfish reached a peak of about eighty-six million (metric) tons, and since that time the global catch has barely been growing. In the mid-1990s, according to the FAO, 74 percent of the world's marine fish stocks were either fully exploited (50 percent), overfished (15 percent), depleted (7 percent), or recovering from past overfishing (2 percent).[5] Not surprisingly, the actual depletion trends vary from place to place. In a few locations, such as the Indian Ocean, the catch continues to grow, while in other locations it is declining steadily. Of the world's fifteen major marine fishing regions, productivity in all but two has fallen, over 30 percent in some fisheries.[6]

In a few locations the stocks have actually collapsed catastrophically. Among the economically most distressing collapses were the California sardine in the 1950s, the Alaskan king crab in the 1960s, and the anchovy off Peru and Chile, which lost about 80 percent of its tonnage in the early 1970s. More recently, stock collapses befell the northern cod off eastern Canada and the cod, haddock, and yellowtail flounder off New England. The economic repercussions, including one hundred thousand lost jobs and millions of threatened livelihoods, have been so severe that the collapses are considered natural disasters.[7]

CAN A TRAGEDY OF THE COMMONS BE AVERTED?

Confronting these sudden declines in fish supplies, one is tempted to view the situation in stark Malthusian terms—a tragedy of the commons with too many boats pursuing too few fish and fishermen bent on exploiting the world's aquatic resources to the point of extinction. It is only a matter of time, according to this pessimistic view, before most of the world's fish stocks will be depleted beyond recovery.

The reality is not as simple, nor as discouraging, as that. Certainly, as mentioned, human activities, including severe economic competition, government subsidies and poor management, have contributed to overfishing. But factors beyond human control have also been at play, including climate cycles affecting ocean temperature, currents, and fish populations, and natural variations in fish production and survival. To reverse the situation and avoid a tragedy of the commons, the fishing community needs to understand better the factors that humans can and cannot control, concentrating its efforts on the former. Fortunately, such an understanding is developing worldwide as marine scientists provide increasingly sophisticated yet practical information about marine ecosystems and resources. Governments and international organizations are also beginning to play a more active and collective role in safeguarding these resources. In any case, the tragedy of the commons is by no means an inevitable outcome: there are many documented examples of human cooperation and collaboration in the use of commonly held resources through which overexploitation was successfully avoided and sustainability achieved.[8]

INFLUENCES OF FISHING ON AQUATIC RESOURCES

A lot more than meets the eye happens as a consequence of fishing. The direct effect, obviously, is to dilute the resource by removing the target fish from their habitat. Loss of the target fish—typically predators at or near the top of a food (trophic) chain—may set in motion a complex set of indirect effects, often cascading events that can alter the character of entire marine communities.[9] One study in the northwest Mediterranean found that the removal of fish (predators) contributes to an increased population of sea urchins (their prey), which in turn depletes the population of edible fleshy algae and leaves crusts of inedible coral-like algae.[10] Such changes can affect the ecosystem's balance of predator–prey populations and ultimately its viability as a source of fishing. Unfortunately, marine ecosystem studies rarely obtain definitive conclusions about fishing impacts, not only because the systems are so complex but also because few baseline studies exist that can define the systems as they were before the onset of human activities. One conclusion is fairly solid, however: the more intensive the fishing, the more serious the indirect ecosystem effects. From an ecosystem perspective, overfishing is a worst-case scenario.

NONHUMAN INFLUENCES ON AQUATIC RESOURCES

Despite the popular notion that overfishing by humans is the sole cause of depleting fish stocks, evidence indicates that ocean ecosystems are influenced as much by natural changes in the physical environment as by human activities.[11] One example is the case of the North Pacific Ocean, where a natural intensification of the Aleutian low-pressure system in 1976 was accompanied by many biological changes, including increased chlorophyl concentrations,[12] increased Alaska salmon catches,[13] and a shift from shrimp to fish dominance in the northern Gulf of Alaska.[14] Another example is the major changes in populations of sardines and anchovy stocks in coastal ecosystems around the world,[15] which are thought to result from long-term, wide-scale changes in physical conditions rather than from human fishing activities.[16]

Recognizing natural changes in fish populations does not lessen the importance of understanding the impacts of human activities but rather points up the need for a balanced view in applying scientific knowledge, with its inevitable uncertainties, to improving resource management. The tendency of some environmentalists and the media to hold humans entirely responsible for observed changes in aquatic resources, when in fact natural events are responsible for many, can lead to misplaced or ineffective policies and actions.

AQUICULTURE

Although over 75 percent of the fish consumed by people still comes from natural marine environments, the portion coming from fish farming, or aquiculture,[17] is increasing rapidly. Aquiculture is in fact the fastest growing sector in world food production, increasing at an annual rate of about 10 percent since 1984, compared with 3 percent for livestock meat and 1.6 percent for capture fisheries production.[18] Aquiculture has become a major contributor to the world's fish supplies: in 1998 the value of aquiculture output reached 25 percent of the world's entire output of fish and shellfish, up from 13 percent in 1990.[19] Asia is the world leader in aquiculture, now producing about 90 percent of the world's aquiculture products, with China contributing about three-quarters of this. Asian aquiculture production dwarfs that of Africa and Latin America, which contribute less than 0.5 percent and 2 percent of global production, respectively.[20] In developing countries carps and tilapias are the most popular aquiculture species, yet more than two hundred species are currently farmed globally in culture facilities as diverse as rice fields, water ponds, and cages and pens; and breeding programs to produce better strains of some species are increasing in number.[21] A considerable share of several high-priced species, including salmon, marine shrimps, and oysters, are produced by aquiculture.

Still, aquiculture has not yet reached the point where it can be considered a major agricultural system along with agronomy and animal husbandry. It must grow to that status if fish farming is to take up the slack between the world's stagnant marine fish production and its growing food requirements. Perhaps even more important, with adequate institutional support aquiculture could contribute significantly to eliminating rural poverty in countries where it is neither a traditional nor widespread practice.[22]

Aquiculture has historically been concentrated in Asian countries, especially China, Indonesia, and Vietnam. More recently a number of internationally sponsored efforts to promote rural aquiculture have taken place in

Thailand, Cambodia, Laos, and Bangladesh. These efforts have concentrated on educating small fish farmers in appropriate technologies for stocking, seeding, fertilizing, and feeding the stocked fish. Initially, most rural fish farmers focus their efforts on subsistence, but even a modest level of success and confidence usually changes their interest to marketing the fish, especially when they begin to produce surpluses beyond their own needs. At this point, with the expectation of higher incomes, many farmers join the market economy. When subsistence farmers choose to pursue aquiculture as a pathway out of poverty, their lack of education and training in pond and stock management techniques can at first cause difficulties and failures. As they gain experience, however, they invariably appreciate the quality of the farm environment as a critical asset for sustainable management and long-term productivity.

In the industrial countries, aquiculture has also been growing significantly, and practitioners range from artisanal fishermen to large industrial-scale multinational companies. In the European Community, species such as trout, salmon, mussels, and oysters remain the staples of aquiculture, but attention is turning increasingly to exotic species that do well in the world market, such as sea bass, sea bream, and turbot. As is the case with other resources, the aquiculture industry in the industrial countries is governed by increasingly strict environmental regulations, to increase food safety and ensure the sustainability of the natural resource base.

FISHING AND THE POOR

A note should be added about the plight of fishers in the poorest countries. In many ways their situation is analogous to the third world's subsistence farmers, whose tragic land degradations were discussed in Chapter 1. A case in point is the overfishing and destruction of coral reefs by many coastal dwellers in the poorest developing countries. As more and more people move to the water's edge, with no source of income other than fishing, they are forced to compete with each other for the small stocks of fish inhabiting the nearby reefs. Thus overfishing is the rule rather than the exception. Some fishers even resort to blowing up the reefs with dynamite to increase their catch. Destruction of the coral reef habitats is clearly very shortsighted as it diminishes and eventually destroys the fish stocks, but people on the edge of starvation are understandably myopic about the benefits of long-term management.

Fishers in the poorest nations, constrained by their immediate need to catch fish for food and livelihood, are often trapped by the vicious cycle of resource overexploitation. In contrast, rich nations have much more flexibility and means to turn around the overfishing problem. They can and must do about overfishing what they have already done to overcome air and water pollution, environmental problems that are mostly under control in the affluent world. With fishing, resource degradation is a more recent problem, and for the vast majority of fisheries it is still quite reversible if intelligent and persistent efforts are made. Many options are open.

HOW TO ACHIEVE SUSTAINABLE AQUATIC RESOURCES

The problems created by overfishing have not yet been solved. Many studies—private, governmental, and international—have concluded that the ways fishing is organized and carried out must change if this precious food resource is to be sustainable in the future. The importance of the problem is widely recognized and the commitment and political will to solve it are growing. Still, no global consensus has yet been reached on the major directions that these changes should take or how they should be implemented. This current lack of consensus does not, however, validate the media's incessant predictions about the demise of the world's fisheries. Nature is much more robust than some would have us believe. Aside from the very few fisheries that the FAO has declared "depleted," the vast majority are capable of regaining their former productivity if the various stakeholders muster enough political will to implement appropriate and scientifically grounded steps. Following are some examples of directions, proposed or already implemented.

Perhaps the most widely supported approach is collective management based on intensive regulation. In the fifteen-nation European Union (EU), a common fisheries policy has been in effect since 1983, dealing with management of fisheries and aquiculture. The approximately 250,000 fishers covered by this policy are licensed and in principle have equal access to member states' waters, except for a coastal band reserved for local fishers. The regulations are pervasive, not only controlling the species and the maximum quantities of fish that may be caught each year but also specifying the maximum time that may be spent fishing and requiring techniques that allow the escape of small fish and reduce capture of by-catch (nontarget) species. The number and sizes of fishing vessels are also subject to control.

The European policy has dealt with the overcapacity issue by requiring member countries to reduce the capacity of their fleets by 30 percent. Thus far few countries have achieved anything near that reduction. The greatest difficulty, however, has attached to the problem of gaining member states' acceptance of the catch quotas and also enforcing those quotas, which are routinely violated by independent fishers. Deep cultural issues also divide member states—for example, dissatisfactions among the Irish people, whose ancestors have fished for thousands of years and who believe that the EU's restrictive catch quotas are inimical to Ireland's socioeconomic and cultural framework.

This characteristically European approach exhibits the well-known limitations and inefficiencies of centralized-government command-and-control regulatory systems and also includes factors unique to the European Community's history and politics. The European fisheries policy has thus far not shown great success, probably because it has done little to alter the fundamental incentives that lead to overfishing. But it is still young and should be given a fair chance to work out its problems and demonstrate whether it can safeguard the future of the great European aquatic commons.

There is another way, more modest but possibly more effective, to avoid the tragedy of the commons. In many places, local fishers manage their own fishing grounds, usually with little government interference, and their management successfully prevents overfishing. For the most part, these arrangements are community based, spontaneously developed, and informally organized. An outstanding example is the Lofoten fishery in Norway, one of the largest commercial cod fisheries in the world.[23] It is totally self-regulated, with no quota regulations, no special licensing system, and no participation by the Norwegian government. The incentive for self-regulation came from problems of crowding and conflicts about fishing gear experienced in the fishery during the latter half of the nineteenth century. The Lofoten fishers realized they needed regulation to solve these problems, but they wanted to carry it out themselves. One hundred years later the system they worked out is still functioning well, and the fishery's cod exports in 1983 were worth over $140 million. The Lofoten system includes fifteen control districts, each with separate, well-defined territories. Each district has committees responsible for developing and implementing its own regulations, enforcing the regulations, and resolving disputes among fishers.[24] Lofoten is an example of Scandinavian pragmatism and cooperation at their finest.

The conditions under which self-regulation of resource commons can be accomplished have been identified by the scholarly work of Elinor Ostrom, who also provides many examples of commons that have been successfully self-managed over long periods without overexploitation.[25] According to Ostrom's findings, self-management has been most successful where physical boundaries were clearly defined, rules were closely linked to local conditions, and sanctions were reliably imposed when rules were broken. Strong community traditions and lack of government interference are also essential factors for self-regulation, both of which are found in the Norwegian fishery example.

A quite different school of thought about fisheries management focuses on increasing the incentives of individual fishers to safeguard the fishing resources that provide their own livelihoods. According to this view, private property rights would be much more efficient than government regulation in providing such incentives. The situation is analogous to property ownership on land: the value of well-kept property tends to rise and that of poorly kept property to fall. If fish stocks were privately owned, owners would be unlikely to rush to take fish nor to deplete their own stocks at the risk of diminishing future catches and endangering their livelihoods. This approach is analogous to the conversion from medieval common ownership of land to the private property system, which is now recognized almost everywhere.[26]

Implementing a private property system for fisheries appears to be most feasible in coastal fisheries, where fish stay put. Coastlines could be divided and private owners allowed to take exclusive possession of the fish in their own areas. Farther from coasts, where most commercially valuable fish are found, it is more difficult to define boundaries and monitor trespass in an area of liquid without obvious property lines. However, new technological developments such as satellite monitoring may enhance the feasibility of assigning rights to such areas.[27]

Although no country has yet completely privatized its fisheries, New Zealand and Iceland have experimented most extensively with property rights management. New Zealand has set "total allowable catch" quotas for the commercial species in each fishing area within its jurisdiction and has sold these quotas to private companies, which can deal with them as divisible and tradable assets. This system appears to have increased aggregate catches and stabilized most fish stocks, while also producing a much longer-term management view of the New Zealand fisheries. Iceland is implementing a similar system that has significantly increased the catches of herring and, to a lesser extent, cod. In response to the quota owners' urging, Iceland has recently reduced the cod quotas, and this already appears to have favorably affected the cod stocks.[28]

SALMON AND DAMS

Whenever two powerful and determined constituencies find themselves on opposite sides of a major environmental conflict, its resolution is likely to be painful and protracted. So it is with salmon and dams. For over six decades a series of major dams built on the Snake River in Washington brought hydroelectric power and the country's cheapest electricity to the U.S. Pacific Northwest. The dams created thousands of jobs and transformed the town of Lewiston, Idaho, into an inland seaport hundreds of miles from the Pacific Ocean. But to the several salmon species native to the Snake River, the dams have been not a blessing but a formidable obstacle to their annual upriver migration to their spawning grounds. Today the river's salmon population has dwindled from hundreds of thousands to perhaps ten thousand, and all species are listed as endangered.

The dams have been the obvious target of blame for the Snake River salmon's near demise, and the federal government has spent over $3 billion in efforts, short of destroying the dams, to save the salmon. None of these has proved effective, and now the previously unthinkable is being considered: breaching the four major Snake River dams, at a cost of another $1 billion. The *New York Times* put it this way: "In the end, the question broadly comes down to whether saving the salmon outweighs the dams' economic benefits to eastern Washington and parts of Idaho."[29] Opposition to removing the dams has, as expected, been intense, and a U.S. senator from Washington stated that breaching the dams would be "an unmitigated disaster and an economic nightmare" for the region.[30] At the time of this writing (April 2001) the dams are still in place.

A research study recently published in the journal *Science* reported evidence that the abundance of Alaska sockeye salmon has fluctuated naturally over the past three hundred years: "Sockeye populations have alternately soared and slipped, following natural climate variations—well before commercial fishers began throwing nets over the sides of boats."[31] Whether recent oceanic temperature changes in the North Pacific Ocean have had any effect on the Snake River salmon population is not known, but the possibility exists that the dams are not the only factor involved in the population decreases. Because of such uncertainties, biologists are not certain that breaching the dams would actually save the salmon.

One thing is certain, though. The magnificent salmon of the U.S. Northwest are considered a precious national asset, and most people do not want them to become extinct. If it is decided, with hard scientific evidence, that human actions have harmed the salmon and that human actions can save them, this affluent nation has the means to save them, and it will do so. In contrast, if this drama were playing out in a country where poverty rules, the salmon would be doomed.

5

IS THE EARTH WARMING?

Is the earth warming? Yes, the earth has warmed since the mid-1800s.[1] Previously, however, the earth had cooled for more than five centuries. Cycles of warming and cooling have, in fact, been part of the earth's natural climate history for millions of years.

If these processes are natural, then what is the global warming debate all about? It is about the proposition that *human use of fossil fuels* has contributed significantly to the past century's warming and that expected future warming may have catastrophic global consequences. However, the evidence for a human contribution is, at best, suggestive. Hard evidence simply doesn't exist. Does that mean that human effects are not occurring? Not necessarily. But media coverage of the global warming issue has been so alarmist that it fails to convey how flimsy the evidence really is. Most people don't realize that many strong statements about a human contribution to global warming are based more on politics than on science.

The climate-change issue has become so highly politicized that its scientific and political aspects are now almost indistinguishable. The United Nations Intergovernmental Panel on Climate Change (IPCC), upon which governments everywhere have depended for the best scientific information, has been transformed from a bona fide effort in international scientific cooperation into what one of its leading participants terms "a hybrid scientific/political organization."[2] And some science policy analysts go further, stating that "previous scholarship has tended to treat the production of scientific knowledge as external to politics.... Science is a human institution deeply engaged in the practice of ordering social and political worlds.... Climate change can no longer be viewed as simply another in a laundry list of environmental issues; rather, it has become a key site in the global transformation of world order."[3] This view would apparently justify the use by scientists/politicians of a legitimate scientific question—global climate change—as a political instrument to create a new, albeit limited, form of world government (i.e., the IPCC) with the power to determine national environmental policy for otherwise sovereign states. I believe that few scientists would accept so radical a view of the role of science in today's world.

Yet, apart from the overheated politics that surrounds the subject of climate change, it remains a fascinating and important scientific subject. Climate dynamics and climate history are extraordinarily complex subjects, and despite intensive study for decades, scientists are not yet able to explain satisfactorily such basic phenomena as extreme weather events (hurricanes, tornadoes, droughts), El Niño variations, historical climate cycles, and trends of atmospheric temperatures. In all these matters the scientific uncertainties are great, and not surprisingly, competent scientists disagree in their interpretations about what is and is not known. In the politicized atmosphere promoted by the IPCC, however, legitimate scientific differences about climate change have become lost in the noise of politics. Even original research results published in prestigious scientific journals often are accompanied by partisan editorial positions. Such politicized science journalism not only confuses scientists and the public but also hinders the objective pursuit of truth through traditional scholarly dialogue.

In recent years the climate-change debate not only has become enmeshed in domestic politics in the industrial countries but has also affected political relations among those countries and with the developing world. One example is the rancorous discord between the United States, European countries, and the developing countries over the Kyoto Climate Change protocol (discussed below).

For some, global warming has become the ultimate symbol of pessimism about the environmental future. Environmental writer Bill McKibben says, "If we had to pick one problem to obsess about over the next fifty years, we'd do well to make it carbon dioxide."[4] Other writers, including myself, believe we'd be far wiser to obsess about poverty than about carbon dioxide.

THE RICH, THE POOR, AND THE CLIMATE

Just how does climate change relate to the subject of this book—poverty, affluence, and the environment? From a scientific focus, the connection arises from the fact that fossil fuels (coal, oil, and natural gas) are the major culprits of the global warming controversy and also happen to be the principal energy sources for both rich and poor countries. From a political focus, the connection arises from the disparity in environmental politics between industrial and developing countries. Governments of the industrial countries have generally accepted the IPCC position that humankind's use of fossil fuels is a major contributor to global warming, and in 1997 they forged an international agreement (the Kyoto protocol) mandating that worldwide fossil-fuel use be drastically reduced as a precaution against future warming.[5] In contrast, the developing nations mostly do not accept global warming as a high-priority issue and, as of this writing, are not subject to the Kyoto agreement. Thus the affluent nations and the developing nations have set themselves on a collision course over environmental policy relating to fossil-fuel use.

Why did a rich-versus-poor political disagreement arise over so fundamental a scientific issue as global climate? How can it be resolved? To gain some insight on these questions, let's have a closer look at the science and politics of global warming. First, the science.

CHEMISTRY 101 AND THE GREENHOUSE EFFECT

The debate about global warming focuses on carbon dioxide, a gaseous substance emitted into the atmosphere when fossil fuels are burned. Environmentalists generally label carbon dioxide as a pollutant; for example, the Sierra Club, in referring to carbon dioxide (see full statement in the Introduction), states: "we are choking our planet in a cloud of this pollution."[6] Introducing the term pollution in this context, however, is misleading since carbon dioxide is neither scientifically nor legally considered a pollutant.[7] Though present in earth's atmosphere in small amounts, carbon dioxide plays an essential role in maintaining life and as part of earth's temperature-control system.

Those who have had the pleasure of an elementary chemistry course will recall that carbon dioxide (CO_2) is one of the two main products of the combustion in air of any fossil fuel (oil, coal, natural gas), the other product being water (H_2O). (Carbon dioxide is also emitted by humans when we "burn" food to create chemical energy.) These combustion products are generally emitted into the atmosphere, no matter whether the combustion takes place in power plants, household gas stoves and heaters, manufacturing facilities, automobiles, or other combustion sources. *The core scientific*

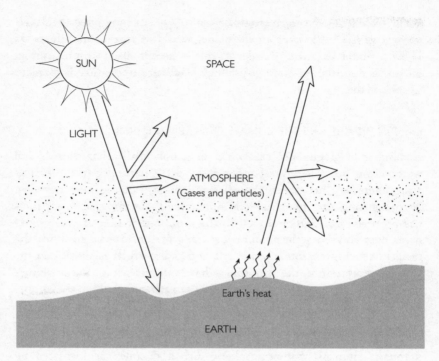

Figure 9.   How the greenhouse effect works. Some of the sun's light (energy) passes through the atmosphere, warming the earth, while some warms the atmosphere and some is reflected back into space. The warm earth gives off heat (infrared radiation), most of which passes through the atmosphere into space, but some is reflected back to earth by molecules in the atmosphere—greenhouse gases, including water vapor, carbon dioxide, methane, and other gases. Tiny particles (aerosols) in the atmosphere have more complicated effects, causing both heating and cooling of the earth and the atmosphere.

issue of the global-warming debate is the extent to which atmospheric carbon dioxide from fossil fuel burning affects global climate.

When residing in the atmosphere, carbon dioxide and water vapor are called "greenhouse gases." They are so named because they trap some of the earth's heat in the same way that the glass canopy of a greenhouse prevents some of its internal heat from escaping, thereby warming the interior of the greenhouse. Figure 9 shows schematically how the greenhouse effect works. By this type of heating, greenhouse gases present naturally in the atmosphere perform a critical function. Indeed, without greenhouse gases the earth would be too cold to have nurtured the development of life.[8] All water on the planet would be frozen, and life as we know it would not exist.[9] Besides its role in greenhouse warming, carbon dioxide is essential for plant physiology. Without carbon dioxide all plant life would die.

CARBON DIOXIDE AND THE CLIMATE CHANGE WHODUNIT

A number of greenhouse gases, including not only carbon dioxide and water vapor but also several other gases, occur naturally in the earth's atmosphere and have been there for millennia. What is new, however, is that during the industrial era, humankind's burning of fossil fuels has been adding *additional* CO_2 to the atmospheric mix of greenhouse gases, over and above the amounts naturally present. Figure 10 shows the results of measurements taken over five decades, which establish that the CO_2 concentration in the atmosphere has been increasing. The preindustrial level of 287 parts per million (ppm) of carbon dioxide in the atmosphere[10] has increased to the current (1998) level of 367 ppm—a 28 percent increase.[11]

This fact is not controversial. Few, if any, scientists question the measurements showing that atmospheric carbon dioxide has increased by almost a third. Nor do most scientists question that humans are the cause of most or all of the carbon dioxide increase. Yet the media continually point to these two facts as the major evidence that *humans are causing the global warming* recently experienced. The weak link in this argument is that empirical science has not established an unambiguous connection between the carbon dioxide increase and the observed global warming. The real scientific controversy about global warming is not about the presence of additional carbon dioxide in the atmosphere from human activities, which is well established, but about *the extent to which that additional carbon dioxide affects climate now or in the future.*

First principles of physics tell us that *some* extra heating must be caused by the additional carbon dioxide in the atmosphere, according to the theory of the greenhouse effect first proposed by the Swedish chemist Svante Arrhenius in 1896.[12] But first principles don't tell us *how much* that heating will be. And first principles do not say *anything whatever* about the possibility of factors other than carbon dioxide that may influence how much the temperature will rise or whether it will rise at all. Besides, and most important, the earth's climate is constantly changing from natural causes, mostly not understood. So the question is: how do

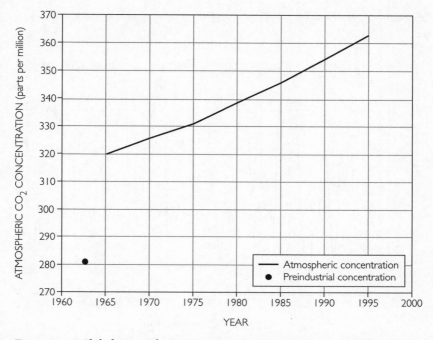

Figure 10.   Global atmospheric CO2 concentration (1965–1996). The carbon dioxide concentration in the atmosphere has increased from the preindustrial level of 280 parts per million (ppm) to 363 ppm at present. Data are from World Resources Institute, *World Resources 1998–1999* (New York: Oxford University Press, 1998).

you distinguish the human contribution, which may be very small, from the natural contribution, which may be either small or large? Put another way, is the additional carbon dioxide humans are adding to the atmosphere likely to have a measurable effect on global temperature, which is in any case changing continuously from natural causes? Or is the temperature effect from the additional carbon dioxide likely to be imperceptible and therefore unimportant as a practical matter?

CLIMATE CYCLES, THEN AND NOW

Keep in mind that global warming is not something that just happened recently. In the earth's long history, climate change is the rule rather than the exception. Studies of the earth's temperature record going back a million years clearly reveal a number of climate cycles, as shown in Figure 11.[13]

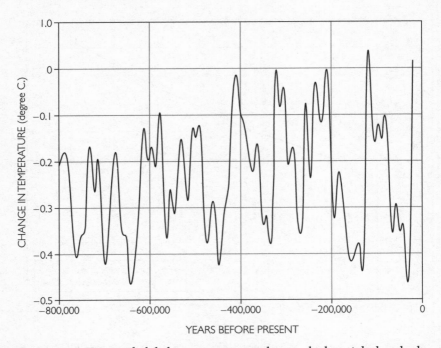

Figure 11. Historical global temperature trends over the last eight hundred thousand years. Temperatures were inferred from oxygen-isotope ratios in sea-floor fossil plankton, based on data from several studies. Graph adapted from T. J. Crowley, "Remembrance of Things Past: Greenhouse Lessons for the Geologic Record," *Consequences* 2 (1996): 3–12.

These cycles have multiple causes—possibly including periodic changes in solar output and variations in the earth's tilt and orbit—but as of now those causes are poorly understood. What we do know for sure is that climate cycles occurred long before humans walked on the planet, and that such warming and cooling trends will continue in the future.

In recent times the earth entered a warming period. From thermometer records we know that the air at the earth's surface warmed about 0.6 degrees Centigrade over the period from the 1860s to the present.[14] The observed warming, however, does not correlate well with the growth in fossil-fuel use during that period. The temperature graph shows that about half of the observed warming took place before 1940, whereas it was only after 1940 that the amounts of greenhouse gases produced by fossil-fuel burning rose rapidly as a result of the heavy industrial expansions of World War II and the postwar boom (80 percent of the carbon dioxide from

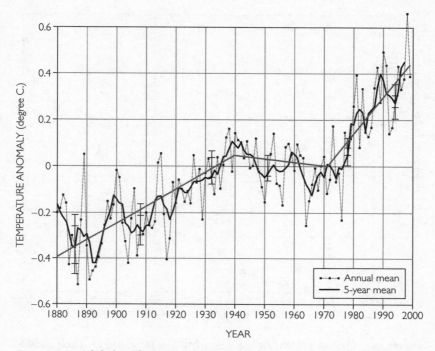

Figure 12.   Global surface air temperatures (1880–2000). Graph from J. R. Hansen, R. Ruedy, J. Glascoe, and M. Sato, "GISS Analysis of Surface Temperature Change," *Journal of Geophysical Research* 104(30) (1999): 997.

human activities was added to the air after 1940).[15] Figure 12 shows, surprisingly, that from about 1940 until about 1980, during a period of rapid increase in fossil-fuel burning, global surface temperatures actually went into a slight *cooling* trend rather than an acceleration of the *warming* trend that would have been expected from greenhouse gases.[16] During the 1970s some scientists became concerned about the possibility of a new ice age from an extended period of global cooling, and several publications reflected that concern, including a report by the prestigious U.S. National Academy of Sciences.[17] Even today, physicist Freeman Dyson expresses the view that "the onset of the next ice age [would be] a far more severe catastrophe than anything associated with warming."[18]

The earth's cooling trend did not continue beyond 1980, but neither has there been an unambiguous warming trend. Since 1980 precise temperature measurements have been made in the earth's atmosphere (troposphere) as well as at the earth's surface, but the results do not agree. The

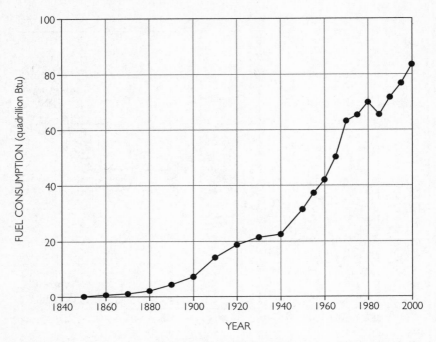

Figure 13.   Historical trend of U.S. fossil-fuel use (1850–2000). Fuels include coal, natural gas, petroleum. Data are from Energy Information Administration, *U.S. Department of Energy Annual Energy Review* (Washington, DC, 2000), tables F1a and F1b.

surface air measurements since 1980 indicate significant warming (0.25 to 0.4 degree Centigrade), while the atmospheric measurements show very little if any warming.[19]

THE TEMPERATURE RECORD

In brief, this is the record: from 1860 to 1940, the earth's *surface* warmed about 0.4 degree Centigrade. Over the six decades since 1940, the earth's *surface* cooled about 0.1 degree Centigrade in the first four decades and warmed about 0.3 degree in the most recent two decades. And in these last two decades, when temperature measurements of the *atmosphere* became available, its temperature has remained essentially unchanged.

Thus, the actual temperature record does not support claims, widely found in the media and environmental literature, that the earth has been steadily warming over the past century. And the observed disparity between the surface and tropospheric temperature trends of the last twenty years has not been explained. A panel of the U.S. National Academy of Sciences examining this disparity observes only that it is "probably at least partially real"[20]—ambiguous language that appears intended to downplay the disparity. Several possible explanations have been offered. First, large urban centers create artificial heating zones ("heat islands") that can contribute to an increase of surface temperature unrelated to greenhouse gases. This could explain why the surface has been heating recently but the atmosphere has not, though one analysis concludes that the heat island effect is too small to fully explain the discrepancy.[21] Second, in the troposphere, soot and dust from volcanic eruptions, such as Mount Pinatubo in 1991, may have contributed to cooling the atmosphere (from blocking the sun's heat), but this cooling should have affected both the surface temperatures and the higher atmospheric temperatures. Also confusing is the fact that the post-1930 surface cooling in the United States far exceeded that of the globe as a whole, and subsequently the surface temperature has reached only the level of the 1930s, despite the presence of large urban areas in the record.[22]

It is frequently claimed that the recent increases in surface temperature are uniquely hazardous to the earth's ecosystems because of the rapidity with which they are occurring—over 0.1 degree Centigrade in a decade. This may be so, but it should be noted that past climate changes were sometimes rapid, as well. For example, around 14,700 years ago, temperatures in Greenland apparently jumped 5 degrees Centigrade in less than twenty years—almost three times the warming from greenhouse gases predicted to occur in this entire century by the most pessimistic scientists.[23]

CLIMATE CHANGE AS AN HISTORICAL FACT

Whatever the present rate of surface warming, there is little justification for the prevalent viewpoint implying that an unchanging climate is the way the earth *ought* to be and that any climate changes now occurring must be caused by humans and should be fixed by humans. In fact, changing climate patterns and cycles have occurred throughout the earth's history. For millions of years, ice sheets regularly waxed and waned as global heating and cooling processes took place.[24] During the most recent ice age, some fifty thousand years ago, ice sheets covered much of North America, Northern Europe, and Northern Asia. Around twelve thousand years ago a warming trend began, signaling the end of the Ice Age and the start of an interglacial period that continues to the present time. This interglacial warm period may have peaked around five thousand to six thousand years ago (the "Climate Optimum"), when global ice melting accelerated and global temperatures became higher than today's. Interglacial periods are thought to persist for about ten thousand years, so the next ice age may be coming soon, that is, in the next five hundred to one thousand years.[25]

Within the current interglacial period, smaller cyclic patterns have emerged. In the most recent millennium, several cycles occurred during which the earth alternately warmed and cooled. There is evidence for an unusually warm period over at least parts of the globe from about 1000 to about 1300, often called the Medieval Warm Period.[26] A mild climate in the Northern Hemisphere during this period probably facilitated the migration of Scandinavian peoples to Greenland and Iceland, as well as their first landing on the North American continent just after 1000. The settlements in Greenland and Iceland thrived for several hundred years but eventually were abandoned when the climate turned colder after about 1450. The cold period, which lasted until the late 1800s, is often called the Little Ice Age.[27] During this cold period agricultural productivity fell, and the mass exodus to North America of many Europeans is attributed at least partly to catastrophic crop failures such as the potato famine in Ireland.

An empirical fact often cited as evidence linking humans and global warming comes from recent indirect studies (such as tree ring records) indicating that the earth became warmer during the twentieth century than at any time in the last four hundred to six hundred years.[28] But this is hardly surprising, since the interval between 1400 and 1880 is known to have been particularly cold (the Little Ice Age referred to above).

A plausible interpretation of most or all of the observed surface warming over the last century is that the earth is in the process of coming out of the Little Ice Age cold cycle that began six hundred years ago. The current warming trend could last for centuries, until the expected arrival of the next ice age, or it could be punctuated by transient warm and cold periods as experienced in the recent millennium.

A HUMAN CONTRIBUTION?

A great deal of global warming rhetoric gives the impression that science has established beyond doubt that the recent warming is mostly due to human activities. That has not been established. While it is possible that human use of fossil fuels could contribute to global warming in the future, there is no hard scientific evidence that it is already doing so. Indeed, the difficulty of establishing a human contribution by empirical observation is formidable. One would need to detect a very small amount of warming caused by human activity in the presence of a much larger background of naturally occurring climate change. There is no way to do this by controlled experiments in a reasonable amount of time, because the "signal" is so weak. It would be like looking for the proverbial needle in a haystack.

Still, understanding climate change is by no means beyond science's reach. Research in the climate sciences is proceeding in several complementary ways. Paleoclimatologists have been probing the earth's past climatic changes and are uncovering exciting new information about the earth's climate history going back thousands and even millions of years. This paleohistory will help eventually to produce a definitive picture of the earth's evolving climate and will in turn help clarify the climate changes we are experiencing in our own era.[29] Such knowledge is indispensable for understanding what the future may hold for the earth's climate. Yet at the present time "we have only scratched the surface of what we need to understand before we can predict our climatic future."[30]

MODELING CLIMATE BY COMPUTER

Mindful of the limited empirical knowledge about climate, some climate scientists have been attempting to understand possible future climate changes by using computer modeling techniques. Global-climate models employ mathematical simulations of the global atmosphere–ocean system based largely on first principles of physics and chemistry. Such models are designed to provide numerical answers to hypothetical questions such as this:

> How much warmer would the earth's atmosphere likely become if fossil-fuel use adds additional carbon dioxide to the atmosphere— for example, twice the preindustrial amount of carbon dioxide?

Powerful computers are used to run the climate models for different assumed values of the heating and cooling forces acting on the atmosphere. (The assumed sets of values are called "scenarios.") The principal heating force included in such calculations is the atmospheric carbon dioxide assumed to come from human fossil-fuel use. For each hypothetical scenario the model computes a value of the atmosphere's temperature response to these physical forces. By running several scenarios, the modelers obtain a set of theoretical projections of how global temperature might change in the future in response to assumed inputs, governed mainly by the levels of fossil-fuel use. But keep in mind that model-generated computer projections are not meant to be *predictions* of the future. They are simply mathematically derived estimates of probable consequences of different assumed initial conditions entered into a computer. Such model results are valuable to climate scientists as tools to help provide a better understanding of the factors that may influence future climate.

Scientists participating in the UN Intergovernmental Panel on Climate Change (IPCC)—the "hybrid scientific/political organization" referred to above—periodically review the results of theoretical projections made from computer-modeling studies. It should be emphasized that, apart from the political misuse of the IPCC's work, some extremely good climate science has been done and continues to be done under its aegis. In its 1996 report, the IPCC concluded that the "most probable" global surface temperature increase from a doubling of atmospheric carbon dioxide by 2100 would be 1.8 degrees Centigrade. Because of uncertainties in the input data and the models themselves, this result was also expressed as a temperature range, from a low of 1.5 degrees Centigrade to a high of 4.5 degrees Centigrade. In the IPCC's updated 2001 report, which employed a larger variety of input assumptions, a "most probable" figure for global temperature rise was not given, and the projected range, for the period 1990 to 2100, was somewhat larger than in the earlier report, 1.4 degrees Centigrade on the low side to 5.8 degrees Centigrade on the high side.[31] The spread in these figures reflects mostly the differences among the models in assumptions about the growth of fossil-fuel use over this century, and differences in how the models handle the climate physics.

These projections reported by the IPCC represent state-of-the-art climate modeling. Yet, like all computer models, they have significant limitations. One problem is that the current models cannot simulate the natural variability of climate over century-long time periods. There are also differences in how various models take account of "feedback" effects, such as the increased water vapor resulting from rising temperatures. A major shortcoming of all current models is that they project only *gradual* climate change, whereas the most serious impacts of climate change could come about from *abrupt* changes. (A simple analogy is the abrupt effects of frost formation, including leaf damage and plant death, when ambient air temperature gradually dips below the freezing point.)

Although the 2001 IPCC summary report highlights its claim that "there is new and stronger evidence that most of the warming observed over the last 50 years is attributable to human activities," no smoking gun unequivocally points to human influence on climate. The strongest indication cited in the report is that "the 1990s have been the warmest decade in the instrumental record, since 1861." Yet that statement refers only to the surface-temperature record and includes no explanation for the disparity with the satellite record, which indicates that the troposphere has barely warmed since 1980. In this matter, the recent National Academy of Sciences report appears to indict the models when it states that a "common aspect [of the models] is the tendency for the lower to mid troposphere to warm more rapidly than the surface."[32] Similarly, Ramanathan and colleagues state that the global greenhouse-gas warming force is expected to be about 40 percent greater in the atmosphere than at the surface.[33] Thus the model outcomes appear to be inconsistent with the observational record. It is also difficult to take seriously the claim that 1998 is "likely to have been the warmest year of the millennium"[34] since no single-year temperature records exist prior to the nineteenth century. In view of the many shortcomings of current climate models, it would be prudent for policy makers to exercise considerable caution about using them as quantitative indicators of future global warming.

THE COMPLICATIONS

The overriding problem with the earlier climate models is that they were based solely on greenhouse gases. But scientists have long been aware that factors other than greenhouse gases can influence atmospheric temperature. Thus the following question:

Are there offsetting factors that could reduce or even reverse the temperature increase caused by human-introduced carbon dioxide?

The answer is yes. A number of physical factors can increase or reduce the warming effects of carbon dioxide. Among the most important are aerosols —tiny particles (sulfates, black carbon, organic compounds, etc.) introduced into the atmosphere by a variety of pollution sources, including automobiles, coal-burning electricity generators, and other industrial sources, as well as by natural sources such as sea spray and desert dust. Atmospheric aerosols are to a large extent products of pollution, not of greenhouse gases. As of now the uncertainties in the total influence of aerosols on climate are large and poorly understood. Some aerosols, such as black carbon, normally contribute to heating the atmosphere because they absorb the sun's heat (though black carbon aerosols residing at high altitudes can actually cool the earth's surface because they block the sun's rays from getting through to the earth's surface). Other aerosols, composed of sulfates and organic compounds, cool the atmosphere because they reflect or scatter the sun's rays away from the earth.[35] Current evidence indicates that aerosols may be responsible for cooling effects at the earth's surface and warming effects in the earth's atmosphere.[36]

At present, the impacts of pollution on the earth's climate are very uncertain. Although the factors involved are difficult to simulate, they need to be fully included in computer models if the models are to be useful as indicators of future climate. When climate models are eventually able to incorporate the full complexity of pollution effects, especially from aerosols, *the projected global temperature change could be either higher or lower than present projections, or could even be zero,* depending on the chemistry of the particular aerosols involved, their altitude, and their geographic region.

In addition to pollution, other physical factors that can influence surface and atmospheric temperature are methane (another greenhouse gas), dust from volcanic activity, and changes in cloud cover, ocean circulation patterns, and air–sea interactions. Changes in the sun's energy output are another possible cause of the earth's warming trend, and these changes correlate well with the twentieth-century temperature pattern.[37] Research is underway on all these physical factors. The complexity of the climate system, however, is such that the uncertainties that compromise climate models may not be significantly reduced in the near future.

In commenting on these complex physical factors, climate expert Benjamin Santer and colleagues said: "There are fundamental and as yet unresolved observational uncertainties, sometimes even in terms of the sign [direction] of the temperature trend."[38] And James Hansen, one of the pioneers of climate-change science, concluded: "The forcings that drive long-term climate change are not known with an accuracy sufficient to define future climate change. Anthropogenic greenhouse gases, which are well measured, cause a strong positive forcing [warming]. But other, poorly measured, anthropogenic forcings, especially changes of atmospheric aerosols, clouds, and land-use patterns, cause a negative forcing that tends to offset greenhouse warming."[39]

Even beyond physical factors, the inherent complexity of the climate system will always be present to thwart attempts to predict future climate. NASA climate scientist David Rind observes: "Climate, like weather, will likely always be complex; determinism in the midst of chaos, unpredictability in the midst of understanding."[40]

KYOTO AND THE NEW CLIMATE POLITICS

Despite the well-recognized scientific limitations of computer climate models, there is a sweeping generality and simplicity in the way the IPCC has presented the computer simulation results that encourages their use by politicians and policy makers. Therein lies the rub: although political negotiating processes have been at the heart of the IPCC's summary reports written for policy makers, the results presented by IPCC have been uncritically accepted in many political circles as representing scientific fact and have become the basis of government climate-change policy throughout most industrial nations. These IPCC model-based projections formed the basis for the 1997 Kyoto agreement, which would require all industrial countries to limit their use of fossil fuels. Massive, costly, and questionable international programs rest on the computer-generated results of this scientific-political organization.

In view of climate's complexity and the limitations of today's climate simulations, one might expect that pronouncements as to human culpability for climate change would be made with considerable circumspection, especially pronouncements made in the name of the scientific community. Thus it was disturbing to many scientists that the 1996 IPCC summary report contained the assertion that "the balance of evidence suggests a discernible climate change due to human activities."[41] The 2001 IPCC revision goes even further, as quoted above, claiming that "there is new and stronger evidence that most of the warming observed over the last 50 years is attributable to human activities." But most of the "new evidence" comes from new computer simulations and does not satisfactorily address either the disparity in the empirical temperature record between surface and atmosphere or the large uncertainties in the contributions of aerosols and other factors. In commenting on the model simulations, the recent National Academy of Sciences report states,

> Because of the large and still uncertain level of natural variability inherent in the climate record and the uncertainties in the time histories of the various forcing agents (and presumably aerosols), a causal linkage between the buildup of greenhouse gases in the atmosphere and the observed climate changes during the 20th century cannot be unequivocally established. The fact that the magnitude of the observed warming is large in comparison to natural variability as simulated in climate models is suggestive of such a linkage, but it does not constitute proof of one because the model simulations could be deficient in natural variability on the decadal to century time scale.[42]

The IPCC reports have been adopted as the centerpiece of most current popularizations of global warming in the media and in the environmental literature. More important, their political impact has been enormous. The 1996 IPCC report became the principal basis for government climate policy in most industrial countries, including the United States. In the report, the IPCC advised that drastic reductions in the burning of fossil fuels would be required to avoid a disastrous global temperature increase. This advice was the driving force behind the adoption in 1997 of the Kyoto protocol, an international agreement aimed at reducing carbon dioxide emissions in the near future. This protocol would require the United States to cut back its fossil-fuel combustion by over 30 percent in order to reach the targeted reduction of carbon dioxide emissions by 2010.

In its original form, the Kyoto protocol had many flaws. First, it exempted from the emission cutbacks developing countries, including China, India, and Brazil, which are increasingly dependent on fossil fuels and whose current greenhouse gas emissions already exceed those of the developed countries.

Second, it mandated short-term reductions in fossil-fuel use to reach the emission targets without regard to the costs of achieving those targets. (In this respect, the Kyoto constraints resemble typical environmental regulation, which often sets quantitative pollution-control targets without considering costs.) A much better approach would be to use cost and price incentives to encourage greatly increased technical efficiency in fossil-fuel use and the development of new energy technologies that produce little or no greenhouse gases. Forced cutbacks in fossil-fuel use could have severe economic consequences for industrial countries and even greater consequences for poor countries should they ultimately agree to be included in the emissions targets. The costs of the cutbacks would have to be paid up front, whereas the assumed benefits would come only many decades later.

Third, the fossil-fuel cutbacks mandated by the Kyoto protocol are too small to be effective. They are considerably more modest than the 60–80 percent cutbacks prescribed by the IPCC; therefore, even if implemented, the Kyoto cutbacks would probably have extremely little if any impact on global climate.[43] By one estimate, only 0.06 degrees Centigrade of global warming would be averted by 2050 from implementing the Kyoto protocol.[44]

The Kyoto protocol was signed in 1997 by many industrial countries, including the United States (Clinton administration).[45] To have legal status, the protocol needs to be ratified by nations that together account for

55 percent of global greenhouse gas emissions. As of June 2002, the protocol has been ratified by 73 countries, including Japan and all fifteen nations of the European Union. Although these countries together account for only 36 percent of emissions, the 55 percent requirement may be met by Russia's expected ratification. Nonetheless, the treaty is unlikely to have real force without ratification by the United States. The current Bush administration opposes Kyoto and has thus far not sought Senate ratification.

Nor did the previous administration seek ratification, despite its signing the initial protocol. Aware that the U.S. Senate had unanimously adopted a resolution rejecting in principle any climate-change treaty that does not include meaningful participation of developing countries,[46] the Clinton administration never brought the Kyoto protocol up for a Senate vote. (The United States had previously ratified the UN Framework Convention on Climate Change, which commits the global community to stabilize CO_2 concentrations "at a level that will prevent dangerous human interference with the climate." But a specific level was not specified in the treaty—and in fact, is not known.)

The new Bush administration actively opposed Kyoto on economic grounds, and the president stated, "Kyoto is, in many ways, unrealistic. Many countries cannot meet their Kyoto targets; the targets themselves were arbitrary and not based upon science. For America, complying with those mandates would have a negative economic impact, with layoffs of workers and price increases for consumers."[47] With the United States retaining its lone dissent, 165 nations agreed in November 2001 to a modified version of Kyoto aimed at easing the task of reducing CO_2 emissions by enabling the international trading of rights to emit CO_2 and also by giving countries credit for expansion of forests and farmland that soak up CO_2 from the atmosphere. The United States did not subscribe to the modified treaty. And some environmental groups deplored the "watering down" of the original Kyoto agreement.

A recent study by economist William Nordhaus assesses the economic ramifications of a Kyoto treaty modified along the above lines. The study finds that the treaty would bring about little progress toward its objective while incurring both substantial costs and political disputes caused by the huge fund transfers resulting from emissions trading. Nordhaus also concludes that U.S. participation in the treaty would have cost approximately $2.3 trillion over the coming decades—over twice the combined cost to all other participants.[48] It does not require sympathy with overall U.S. climate-change policy to understand the United States' reluctance to become such an unequal partner in the Kyoto enterprise.

Although the political controversy over carbon dioxide emissions continues unabated, the scientific status of the climate-change problem has evolved. The science has moved away from its earlier narrow focus on carbon dioxide as a predictor of global warming to an increasing realization that the world's future climate is likely to be determined by a changing mix of complex and countervailing factors, many of which are not under human control and all of which are poorly understood. With persistent research, a more mature science of climate and climate change will evolve and will guide the world's peoples in developing climate policies that take better account of the needs of both affluent and developing countries.

HOW MUCH DOES GLOBAL WARMING MATTER?

Regardless of the causes, we do know that the earth's surface has warmed during the past century. Although we do not know the extent to which it will warm in the future or whether it will warm at all, we need to ask this critical follow-up question: How much does global warming matter?

What would be the consequences to society if the global average temperature did actually rise during the current century, say, by about two degrees Centigrade?

Some environmentalists have predicted dire consequences, including severe weather extremes, loss of agricultural productivity, rise in sea level with destruction of coastal and island environments, and spread of diseases. Activists press for international commitments much stronger than Kyoto to reduce humankind's combustion of fossil fuels, justifying these as precautionary measures ("insurance") in case the most pessimistic predictions of global warming turn out to be correct. Others counter that this cure would be worse than the disease—that is, the social and economic impacts of proposed government sanctions forcing reductions in fossil-fuel use would be more serious than the effects of a temperature rise, which could be small or even beneficial.

Although the debate over human impacts on climate probably won't be resolved for decades, a case can be made for adopting a less alarmist view of a warmer world. In any case, the warmer world is already here. Look at the historical evidence on the effects of temperature change. In the last twenty-five hundred years, global temperatures have varied by more than three degrees Centigrade, and some of the changes have been much more abrupt than the gradual changes projected by the IPCC.[49] During all of recorded history humans have survived and prospered in climate zones that differ from each other far more than the changes in global temperatures now being discussed. Today people show a definite preference for warmer climates. In the United States, one of the few places where environmental migration is possible within the same political entity, the migration from the cold Northeast to the warm Southwest is far greater than the reverse flow.

Those who predict agricultural losses from a warmer climate have most likely got it backwards. Warm periods have historically benefited the development of civilization, and cold periods have been detrimental. For example, the Medieval Warm Period, from about 900 to 1300, facilitated the Viking settlements of Iceland and Greenland, whereas the subsequent Little Ice Age led to crop failures, famines, and disease. Even a small temperature increase brings a longer and more frost-free growing season—a definite advantage for many farmers, especially those in large cold countries such as Russia and Canada.[50] Enrichment of atmospheric CO_2 is well-known by agronomists to stimulate plant growth and development in greenhouses; thus such enrichment at the global level can be expected to lead to an increase in global vegetative or biological productivity, as well as an increase in water-use efficiency.[51] Since a warmer climate resulting from atmospheric CO_2 enrichment would improve plants as biological converters of solar energy, it is not unreasonable to surmise that global warming could be beneficial to agriculture. Studies of this issue from an economic perspective have reached the same conclusion: moderate global warming would most likely produce net economic benefits, raising gross national product and average income, especially for the agriculture and forestry sectors.[52] Such projections of the future, of course, are subject to great uncertainty and cannot exclude the possibility that unexpected negative impacts would occur.

Concerns have been raised that warmer temperatures would spread insect-borne diseases such as malaria, dengue fever, and yellow fever through increased precipitation leading to expansion of favorable habitats.[53] There is no solid evidence for this concern. These illnesses were common in North America, Western Europe, and Russia during the nineteenth century, when the world was colder than it is today. Although the spread of disease is a complex matter, the main carriers of these diseases are most likely humans traveling the globe and insects traveling with people and goods. The main allies against future disease are surely not cold climates but rather improvements in regional insect control, water quality, and public health. As poverty recedes and people's living conditions improve in the developing world, the level of disease, and its spread, can be expected to decrease.[54] Dr. Paul Reiter, a specialist in insect-borne diseases, puts it this way:

> Developed countries like the United States need not fear the spread of insect-borne diseases provided they remain prosperous. Insect-borne diseases are not diseases of climate but of poverty. Whatever the climate, developing countries will remain at risk until they acquire window screens, air conditioning, modern medicine, and other amenities most Americans take for granted. As a matter of social policy, the best precaution is to improve living standards in general and health infrastructures in particular.[55]

One of the direst (and most highly publicized) predictions of global warming theorists is that greenhouse gas warming will cause sea level to rise and that as a result many oceanic islands and lowland areas, such as Bangladesh, may be submerged.[56] But in fact, sea level is rising now and has been rising for thousands of years, once having been low enough to expose a land bridge between Siberia and Alaska over which humans walked in their migrations from Asia to North America. Recent analyses suggest that sea level rose at a rate of about one to two centimeters per century (0.4 to 0.8 inches) over the last three thousand years.[57] Direct sea-level measurements made throughout the twentieth century have been interpreted in some studies to show that the level is presently rising at a much faster rate, about ten to twenty-five centimeters per century (4 to 10 inches),[58] but other studies conclude that the rate is much lower than this.[59] To whatever extent sea-level rise may have accelerated, the change is thought to have taken place before the period of industrialization.[60]

The question is, of course, whether the ongoing sea-level rise has anything to do with human use of fossil fuels. Before looking at that, however, let's take a step back and ask what science has to say about how global temperature change may relate to sea level change. This is more complicated than it first appears. One factor is that water expands as it warms, which would contribute to *rising* sea level. A factor that could work in the opposite direction is that warming increases evaporation of ocean water, which could increase the snowfall on the Arctic and Antarctic ice sheets, removing water from the ocean and *lowering* sea level. The relative importance of these two factors is not known. We do know from studies of the West Antarctic Ice Sheet that this ice sheet has been melting continuously since the last great ice age, about twenty thousand years ago, and that sea level has been rising ever since.[61] Continued melting of this ice sheet until the next ice age may be inevitable, in which case sea level would rise by fifteen to eighteen feet when the sheet is completely melted. Other mechanisms have been suggested for natural sea-level rise, including tectonic changes in the shape of the ocean basins.[62] The theoretical computer climate models attribute most of the sea-level rise to thermal expansion of the oceans, and thus they predict that further global temperature increase (presumably from human activities) will accelerate the ongoing sea-level rise. Since these models are unable, however, to deal adequately with the totality of natural phenomena involved, their predictions about sea-level rise should be viewed skeptically.

The natural causes of sea-level rise, such as those just mentioned, are part of the earth's evolution. They have nothing to do with human activities, and there is nothing that humans can do about them. Civilization has always adjusted to such changes just as it has adjusted to earthquakes and other natural phenomena over which humans have no control. This is not to say that adjusting to natural changes is not sometimes painful; certainly, adjusting to earthquakes and tornadoes is very painful. But if there is nothing we can do about certain natural phenomena, we do adjust, whether it is painful or not. Sea-level rise is most likely a phenomenon over which humans have no control.

As to the unfortunate flood victims in vulnerable low-lying areas such as Bangladesh, they should be assisted by the international community out of humanitarian considerations regardless of the causes of their frequent flood disasters. Such assistance should in no way be tied to the vicissitudes of political or scientific debate over complex subjects such as global warming or ice-sheet melting.

Another claim of some environmentalists is that weather-related natural disasters, including hurricanes, tornadoes, droughts, and floods, have been increasing in frequency and severity, presumably as the result of human-caused global warming. The actual historical record does not support such claims. On the contrary, several recent statistical studies have found that natural disasters, including hurricanes, typhoons, tropical storms, floods, blizzards, wildfires, heat waves, and earthquakes, have not been increasing in frequency.[63] The *costs* of losses from natural disasters are indeed rising, to the dismay of insurance companies and government emergency agencies, but this is because people in the affluent societies have increasingly constructed expensive properties in areas vulnerable to natural hazards, such as coastlines, steep hills, and forested areas.[64] They continue to do so not only because such areas often provide the most attractive sites for habitation but also because the costs of disaster insurance are spread over the larger society and thus are relatively low to the insured.

Since society has choices, one should ask what would be the likely effects if, on one hand, people decided to adjust to climate change, regardless of its causes, or on the other, governments implemented drastic policies to attempt to lessen the presumed human contribution to the change. At least from an economic perspective, adjusting to the change would almost surely be the winner. Several analyses have projected that the overall cost of the worst-case consequences of warming would be no more than about a 2 percent reduction in world output.[65] Since average per-capita income will probably quadruple during the next century, the potential loss seems small indeed. A more recent economic study emphasizing adaptation to climate change indicates that in the market economy of the United States the overall impacts of modest global warming are likely to be beneficial rather than damaging. The amount of net benefit is small, about 0.2 percent of the economy.[66] (One must always keep in mind the statistical uncertainties inherent in such analyses; i.e., there are small probabilities that the benefits or costs could turn out to be much greater than or much less than the most probable outcomes.)

In contrast, the economic costs of governmental actions ("insurance" policies) restricting the use of fossil fuels could be large. I quoted above the recent Nordhaus study projecting a $2.3 trillion cost to the U.S. economy over coming decades from compliance with the Kyoto treaty.[67] One U.S. government study suggested that a cost-effective way of bringing about fossil-fuel reductions would be a combination of carbon taxes and international trading in emissions rights.[68] Emissions rights trading was in fact included in the modified Kyoto agreement. Such trading schemes would result in huge income transfers as rich nations pay poor nations for emissions quotas that the latter would probably not have used anyway. It is not reasonable to assume that the rich nations would be willing to do this.[69]

Taking into account the large uncertainties in estimating the future growth of the world economy and corresponding growth in fossil-fuel use, one group of economists estimates that the costs of greenhouse-gas reduction would be in the neighborhood of 1 percent of world output,[70] while another estimate is higher, around 5 percent of output.[71] The costs would be expected to be considerably higher if large reductions were forced upon the global economy over a short time period or if the most economically efficient schemes to bring about the reductions were not actually employed —a likely possibility. Political economists Jacoby, Prinn, and Schmalensee put it more strongly: "It will be nearly impossible to slow climate warming [sic] appreciably without condemning much of the world to poverty, unless energy sources that emit little or no carbon dioxide become competitive with conventional fossil fuels."[72]

So what is my bottom line about global warming? First, some warming has been underway for over a century, at least partly from natural causes, and the world has been adjusting to it as it did with past climate changes. Second, if it turns out that human activity is adding to the natural warming, the amount will probably be small, and society can adjust to that as well, at relatively low cost or even net benefit. Third, the industrial nations are not likely to carry out inefficient, Kyoto-type mandated reductions in fossil-fuel use on the basis of so incomplete a scientific foundation as presently exists. The costs of following the "precautionary principle" in this way could well exceed the potential benefits. Far more effective would be policies and actions by the industrial countries to accelerate the development, in the near term, of technologies that utilize fossil fuels (and all resources) more efficiently and, in the longer term, technologies that do not require use of fossil fuels.

Finally, the industrial nations should ensure the future credibility of climate science by totally separating the pursuit of this important science from global politics. The affluent countries should continue to support strong climate-research programs, which will improve the theoretical understanding of and empirical database on factors that influence long-term climate change, and also increase understanding of short-term weather dynamics. Such research not only is relevant to the greenhouse gas issue but also will richly reward humankind by improving people's ability to cope with extreme weather events such as hurricanes, tornadoes, and floods, whatever their causes.

6

WATER, WATER EVERYWHERE

In his book *Tapped Out*, former senator Paul Simon writes, "It is no exaggeration to say that the conflict between humanity's growing thirst and the projected supply of usable, potable water could result in the most devastating natural disaster since history has been recorded accurately, unless something happens to stop it." Citing the statistic "per capita water consumption is rising twice as fast as the world's population," Simon states, "You do not have to be an Einstein to understand that we are headed toward a potential calamity."[1] With this classic example of environmental pessimism as a backdrop, let us look at some facts about water.

Water is one of the earth's most critical resources. Like air, it is essential to support life. The earth has basically two sources of potable (drinkable) water: freshwater and groundwater. (Salt water from the oceans, although unlimited in amount, is not potable unless the salt is removed, a very expensive process.) Freshwater comes from precipitation—rain, snow, and sleet. Although freshwater is a truly renewable resource, its replenishment depends on annual precipitation, which is not only bounded in amount but also varies from year to year.

Groundwater is more complicated. Found underground almost everywhere in rock crevices and cracks, and in spaces between rocks and soil, groundwater can be thought of as a hidden underground lake whose surface is called the "water table."[2] The groundwater "lake" is not a single pool but a series of interconnected water-bearing formations. The groundwater resource is neither nonrenewable nor completely renewable. Although it is continually replenished from precipitation seeping into the ground, the replenishment rate can be slow, sometimes much slower than the rate at which water is pumped out for human uses. When groundwater is over-pumped, not only will the resource be depleted, but also its quality can be degraded by intrusion of salt water from the sea. The quality of ground-water resources can also be degraded by contamination from human activities such as waste disposal sites or chemical storage tanks. If they are to be sustainable over time, groundwater resources must be carefully managed and protected from contamination and overuse.

Freshwater is a plentiful and renewable resource. A quick look at the numbers shows that the earth's total supply of freshwater is extremely large, although only a small fraction is readily usable. Over the globe, the annual total rain and snowfall averages about 577,000 cubic kilometers (ckm), of which 119,000 ckm falls over land.[3] Of the latter amount, about 72,000 ckm evaporates, leaving 47,000 ckm for surface water runoff and groundwater recharges. Much of this runoff, however, is inaccessible or inconvenient for human withdrawal; one estimate is that only 12,500 ckm qualify as "accessible runoff."[4] In 1900 about 580 ckm of freshwater were actually used globally, about 4.6 percent of the amount now considered readily accessible. By 2000 the use had grown to 4,000 ckm,[5] about 32 percent of the readily accessible amount. The most recent forecasts of global water use in the year 2025 range from about 3,600 to 5,500 ckm, representing 29–44 percent of the currently accessible fresh water.[6]

Do these figures support Senator Simon's pessimistic conclusion that the world is headed for a water calamity? Probably not. The annual global supply of freshwater is more than adequate to serve the eight to nine billion people expected to inhabit the earth during this century.[7] A water calamity could occur, of course, if people were to revert to traditional assumptions about water—that water is a free good, that its supply is essentially infinite, that the water will be there no matter how we treat, or mistreat, the world's water supplies. But evidence suggests that civilization no longer thinks about water in that way. People everywhere recognize that there is no substitute for freshwater, that civilization must safeguard the earth's water resources.

Because water is so basic a resource, it poses many issues for society—issues of supply, distribution, cost, quality. As with other resources, the dimensions of these water problems are quite different in rich and poor countries. The water policies and practices of the affluent societies have been moving in the right direction for some time. In the United Kingdom, for example, the government has established national priorities for groundwater management "to protect a priceless national asset."[8] Most other affluent countries have set similar priorities and have seen major improvements in their water systems. In contrast, among the world's poorest countries water supply and quality problems have worsened and in some cases have become a limiting factor in their economic and social development. It is estimated that in developing countries only 30 to 40 percent of the people have access to freshwater, and a much lower percentage have access to potable water. These problems are due less to the intrinsic challenges of the water resource than to the intrinsic challenges of poverty.

In this chapter we'll cover three major water issues: distribution of freshwater resources, efficiency of water use, and water quality and public health.

DISTRIBUTION OF FRESHWATER RESOURCES

Overall there is more than enough freshwater to go around. Unfortunately, freshwater resources are distributed very unevenly around the globe, and much of the freshwater is not located where the people who want to use it are located. Some countries are hugely endowed with freshwater, while others have practically none. A useful measure of the renewable freshwater supply is the annual amount available *per person*, a quantity that varies enormously from country to country. A sampling of current data shows that sparsely populated Iceland, with 606,500 cubic meters of freshwater per person, is by far the world's water-richest country, followed by Surinam (453,000) and Guyana (282,000). At the other extreme, Kuwait, with only 11 cubic meters per person, Egypt (43), and United Arab Emirates (64) are the water-poorest countries. Canada (94,000), Norway (88,000), Russia (29,000), Sweden (20,000), and the United States (8,900) are well endowed with domestic water supplies. Italy (2,800), China (2,200), and the United Kingdom (1,200) are less well endowed, while Belgium (822), the Netherlands (635), and Israel (289) are much less well endowed.[9] As a rule of thumb, any country with available freshwater resources of 1,000–1,600 cubic meters per person per year faces water stress and can expect to suffer water shortages at certain times and places. In countries with less than 1,000 cubic meters per person per year, water availability is considered a severe constraint on socioeconomic development and environmental quality.[10] It has been estimated that about 8 percent of the world's population lives in countries that are highly water stressed.[11]

In some water-poor countries the annual demand for freshwater is much greater than the amount replenished each year. In such circum-

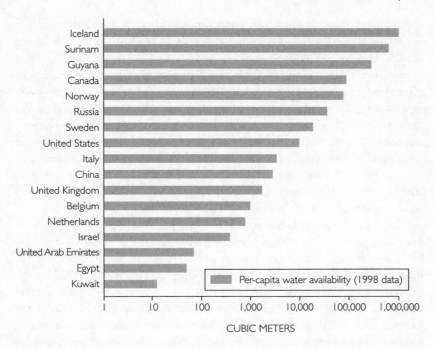

Figure 14. Per-capita annual freshwater availability, selected countries. Note the scale's wide range, going from 1 to 1 million cubic meters. Data are from World Resources Institute, 1998–1999 *World Resources* (Oxford, UK: Oxford University Press, 1998), table 12.1.

stances water shortages are chronic, and even one season of less than normal rainfall can be disastrous. Countries in this situation often resort to drawing down nonrenewable groundwater sources or drawing water from inflowing rivers, sometimes seriously depleting both their flow and their fish populations. Among the most water-stressed countries, Kuwait annually uses twenty-seven times its renewable water resource, Egypt uses twenty times, and the United Arab Emirates fourteen times. Less extreme but still highly water-stressed are Israel, which uses 109 percent of its available renewable supply, Belgium (108 percent), and the Netherlands (78 percent). At the other extreme, water-rich Norway, Sweden, and Canada use less than 2 percent of their available renewable supply. In between are countries under low to moderate water stress, such as China, which uses 16 percent of available supply; the UK, 17 percent; the United States, 19 percent; and Italy, 35 percent. (These figures are approximate and should be used only for rough comparisons among countries.)

Countries that have brackish water supplies or access to saltwater bodies can in principle produce unlimited amounts of freshwater by employing desalination technologies to remove the salt. A wide range of technologies is available, from multistage flash distillation to high-tech applications of electrodialysis and reverse osmosis. Despite years of research and $1.5 billion of research and development investment by the U.S. government, the desalination technologies currently available are very expensive and energy consuming, so that beyond desalting small amounts of water for specialized uses, desalination is practical only for wealthy and energy-rich countries. In this category are Saudi Arabia, the world leader in desalination (5.4 million cubic meters per day), followed by the United States (3.6 million), United Arab Emirates (2.2 million), and Kuwait (1.6 million).[12]

The statistics quoted above point to some places where water shortages are likely to occur even though the overall global water resource is plentiful. As with most resources, however, the impacts of water shortages are dramatically different in rich and poor countries. This is illustrated by the experience of California during the drought of 1987–1993.[13] In normal years California's average rainfall is only about one-quarter of that of the eastern United States, though the state's average water consumption is much higher. During the drought years, less than normal precipitation occurred, and the water level of many of the state's reservoirs fell to less than a third of capacity. During that period California's cities rationed water, and the state agencies that control California's water distribution system set up a water bank and drastically cut farmers' supplies, leading to the closure of some highly productive vegetable farms. In order to survive the drought, many farmers pumped water from already depleted groundwater supplies. Although the economic and environmental losses were great, the situation never reached emergency proportions, and California recovered quickly once precipitation returned to normal levels.

Contrast this with the humanitarian emergency that developed in Ethiopia in 1999, when inadequate rainfall occurred for several years. The result was severe shortages of food and medicine; an alarming increase in deaths from malaria, measles, and diarrhea among young children weakened by malnutrition; and extensive livestock disease and death. Or consider the drought emergency in Tanzania in 2000, when generalized crop failures followed a season of scarce, erratically distributed rains, and three million Tanzanians faced famine and starvation until food shortages were relieved by international emergency food distributions.

Clearly, the vagaries of freshwater supply, although responsible for triggering the onset of drought conditions in many countries, are not the fundamental cause of the misery that can accompany drought. Poverty is the major cause. Lacking sufficient financial resources and technical expertise for water management, countries with low per-capita incomes are in a poor position to respond to water scarcity. Subsistence farmers in such countries have been chronically unable to grow or otherwise obtain enough food to provide a cushion against poor harvests during the occasional dry period. Another contributing factor in many poor countries is their woeful lack of institutions dedicated to rapid emergency assistance.

Throughout history, tight water supplies have been the cause of many tensions and conflicts between groups and nations. These conflicts can be especially sharp when the water supplies are shared, as, for example, in the over 220 river basins that now traverse two or more countries.[14] The water-shy Middle East has been the site of water disputes for millennia, and in recent times a number of political conflicts in that region have been exacerbated by water issues. In the 1960s, for example, Syria began operations to divert the headwaters of the Jordan River away from Israel, which responded by taking military action against the diversion facilities.[15] During the 1967 Arab-Israeli War Israel won control of all the headwaters of the Jordan as well as the groundwater of the West Bank. As of 2000 one of the main impediments to progress in peace negotiations between Syria and Israel is their continuing disagreement over details of the boundaries between the two countries. The areas under dispute are small, but the impact on control of the Jordan River headwaters is huge.

Fortunately cooperation is gradually replacing conflict as countries around the world realize that partnership in the shared development of water resources can be mutually advantageous for all parties. The following principles were enunciated at an international water conference in Dublin in 1992.

- Freshwater is a finite and vulnerable resource, essential to sustain life, development, and the environment.

- Water development and management should be based on a participatory approach, involving users, planners, and policy makers at all levels.

- Women play a central part in the provision, management, and safeguarding of water.

- Water has an economic value in all its competing uses and should be recognized as an economic good.[16]

These principles are actually being applied in resolving water conflicts. An outstanding example is the 1994 peace treaty between Israel and Jordan, which explicitly resolved a number of contentious water issues over the Jordan River basin and included a comprehensive plan for water allocations, data sharing, and joint management and development of the scarce Jordan River water supply. Jordan's enhanced confidence in the water supply arising from this agreement helped stimulate its government to make a number of changes in its own water sector, including privatizing Amman's supply and distribution system, improving the system technically to reduce water losses, and altering the price structure to reflect water's true value.

Such agreements help build confidence that water conflicts around the world, often stimulated by political differences rather than by intrinsic resource scarcity, can be mitigated by sincere negotiations among the involved parties. But much remains to be done, as, for example, in the unresolved disputes between Syria, Iraq, and Turkey over water quantity and quality in the Tigris and Euphrates river systems. One can conclude that if a "water calamity" ever does occur, its cause will more likely be a shortage of political will than a shortage of water.

EFFICIENCY OF WATER USE

Granted that humans cannot do much to alter the uneven distribution of the world's rainfall, the fact is that local and regional water shortages are caused not only by uneven water availability but also because people simply use too much water. Very few places in the world are exempt from the problem of people using too much water. The good news is that the problem is widely recognized and that worldwide efforts to improve efficiency, in rich and poor countries alike, have the promise of bringing vast new water "supplies" to the world. Nonetheless, as with other resource issues, solving the water inefficiency problem is proving to be far more difficult in poor countries than in rich countries.

Although urban areas and industrial enterprises worldwide are guilty of using water inefficiently, agriculture is the major culprit. Indeed, the greatest challenge to the adequacy of the world's future water supply comes from agriculture. Overall, agriculture draws about 75 percent of the world's water, and in some places the fraction is considerably higher; for example, in Africa agriculture may use as much as 88 percent of the continent's water.[17] Yet the overall efficiency of agricultural water use worldwide is only about 40 percent.[18] This implies that more than half of all water used in agriculture never contributes to food production. Even small improvements in agricultural water efficiency can have large impacts because of the dominance of agriculture in the world's water economy. Such improvements are essential both to maintain growth in agricultural productivity without additional sources of water and to allow more water to be reallocated from agriculture to urban and industrial uses.

Irrigation is the sine qua non of productive agriculture. Although only about 18 percent of the world's cropland is irrigated, irrigated land accounts for about 40 percent of world crop production[19] and two-thirds of rice and wheat production.[20] Inefficient irrigation practices on cropland are responsible for most water losses in agriculture. Water is lost to the ground as it passes through leaky irrigation pipes and unlined aqueducts, and water evaporates from surface canal systems, irrigation furrows, and flooded fields. Such losses can be drastically reduced by switching to modern irrigation methods, including sprinkler and drip irrigation systems that allow water to be delivered precisely when and where it is needed. These new irrigation methods are being increasingly applied in the affluent countries, although the changeover proceeds slowly. At the beginning of the 1990s, drip irrigation was used on only 6 percent of agricultural land in Australia, while 13 percent of California land was drip irrigated.[21]

The main impediment to increasing the efficiency of water use, whether in irrigation or other uses, is underpricing. This is the case in both rich and poor countries. In the United States, farmers typically pay only one-fifth of the true cost of irrigation from federal reclamation projects.[22] In California, the leading U.S. agricultural state, inefficient use of agricultural water continues to be encouraged by government allocation programs that heavily subsidize the price. Agribusinesses in California's Central Valley pay as little as $9 per acre-foot, while residents of the coastal city of Santa Barbara are prepared to pay as much as $2,000 per acre-foot for desalted water from a plant built as a backstop against water emergencies. The extremely low cost of agricultural water in California, as in other western states, encourages the production of a range of crops that are both low valued and highly water consumptive. Four crops—rice, cotton, alfalfa, and irrigated pasture—consume 57 percent of California's agricultural water but produce only 17 percent of its agricultural revenue.[23] Enough water to meet the entire needs of Los Angeles's thirteen million residents is being used in California to grow irrigated pasture for livestock, although the pasture's economic value is only about 0.03 percent the value of the Los Angeles region's $300 billion economy.[24]

When water is available at low cost because of subsidies, there is little incentive to improve either physical efficiency (e.g., through better piping) or economic efficiency. Underpriced water encourages wasteful use not only in agriculture but also in urban areas. Paradoxically, underpricing creates shortages that the rich meet by buying water from private vendors at outrageous prices, leaving poor folk at the mercy of extortionists or without water. In the developing world, underpricing is rampant. For example, city dwellers in Indonesia and Pakistan pay only around 5 percent as much for piped water as do their counterparts in Germany, and water systems in the developing world often suffer from 40-60 percent losses in providing municipal services.[25] In the United States too, some cities do not even meter the usage of water, thus providing households with no incentive to use water efficiently. In the U.S. Southwest, aquifers are being dangerously overdrawn while subsidized water makes the deserts bloom and lush green urban lawns mimic those of the rain-rich U.S. Northeast. Water systems continue to leak and faucets continue to drip, but consumers are not likely to be concerned so long as the water is cheap and keeps coming.

But the water doesn't necessarily keep coming. In some places supplies of water are chronically limited (as the statistical data on previous pages show), and in other places supplies become limited during prolonged dry periods. If governments keep water prices artificially low so that consumers' use patterns continue to be inefficient, the inevitable result will be shortages, often followed by rationing or other imposed restraints on demand. In such situations governments may be unable to provide additional supplies at any cost to reduce or eliminate the supply–demand gap.

The traditional view of water as an entitlement rather than an economic good is deeply ingrained in most societies, and governments throughout the world have reinforced that view by subsidizing the price of water. But the entitlement view of water is now changing, as evidenced by the 1992 Dublin principles, one of which states, "Water has an economic value in all its competing uses and should be recognized as an economic good." Recent trends in market pricing of water are bringing increased efficiency of use. For example, a study of twenty-three U.S. cities showed that a 10 percent increase in the price of water would reduce water consumption between 3.8 and 12.6 percent.[26] Similar estimates for the agricultural sector note that, starting from a price of $17 per acre-foot, a 10 percent increase in price would yield a 20 percent decrease in water use in California.[27]

To achieve continuing increases in water use efficiency, governments need to move away from centralized allocation schemes and toward policies that allow the price of water to reflect its true value, to all users. Many economists believe that a market system allowing farmers to sell surplus water would stimulate higher-valued uses and would improve the overall efficiency of water use. The good news is that such market-directed

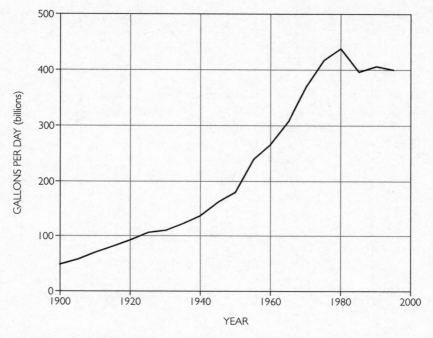

Figure 15.   Trend in total U.S. water use (1900–1995). Data from 1900 to 1950 are taken from Peter H. Gleick, *The World's Water* (Washington, DC: Island Press, 1998), 245, table 3. Data from 1950 to 1995 are taken from Wayne B. Solley, Robert R. Pierce, and Howard A. Perlman, *Estimated Use of Water in the United States in 1995*, U.S. Geological Survey circular 1200 (Washington, DC, 1998).

changes in water policy are increasingly being considered or implemented, not only in the United States but around the world, and the prognosis is excellent for continuing improvements in water use efficiency.

The new price trends and water policies have already had an important positive impact on the efficiency of water use in the United States. From 1950 to 1975 total U.S. off-stream water use (i.e., withdrawals) grew at a rate of about 2.8 percent per year. Had this growth rate continued, U.S. water use would double every twenty-five years. It would not take many such doublings to produce severe water shortages in the country. However, in spite of growing population, total U.S. water consumption actually leveled off in the late 1970s and has been declining ever since. Total U.S. water consumption was in fact 2 percent less in 1995 than in 1990 and 10 percent less than in 1980. And per-capita use—the all-important measure of water

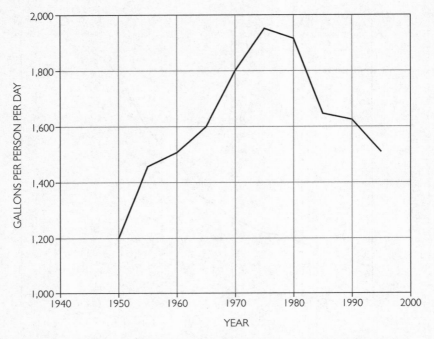

Figure 16.   Trend in per-capita U.S. water use (1950–1995). Data are from Wayne B. Solley, Robert R. Pierce, and Howard A. Perlman, *Estimated Use of Water in the United States in 1995*, U.S. Geological Survey circular 1200 (Washington, DC, 1998).

use efficiency—declined in the United States by more than 20 percent between 1980 and 1995.[28]

The recent improvement in efficiency of U.S. water use demonstrates that the United States, while only moderately endowed with water resources, does not have an overall water-supply problem. Looking to the future, water expert Peter Gleick argues that development of new sources of water supply in the United States can largely be avoided if several things happen:

· Implementation of intelligent water conservation and demand-management programs

· Installation of efficient new equipment

· Application of appropriate economic and institutional incentives to shift water among users[29]

The record of steadily increasing water-use efficiency in the United States supports the notion that affluent societies have both the will and capability of sustaining their precious resources. And with a reasonable expectation of continuing technological innovations and increasing reliance on market forces, optimism is warranted about the future of water supplies everywhere.

One caveat should be added, however, to observations about the importance of market forces in developing and maintaining an adequate water supply. Although the market is indeed important, equally important is the notion that not all of the values that water brings to human society can be captured in traditional monetary terms. Affluent societies may choose, for example, to make substantial investments to preserve certain ecosystems and ecological cycles for the future even though the scientific and economic advantages of such actions cannot be convincingly demonstrated now. Such policies embody social choices that affluent democratic societies are able to make. Poverty-stricken societies do not have the luxury of such choices.

WATER QUALITY AND PUBLIC HEALTH

In the poor countries, lack of access to water is a major problem for many people, as discussed above. But a far more serious problem is lack of access to basic sanitation. According to the World Health Organization, 2.6 billion people lacked basic sanitation services in 1990, and 1.3 billion were without access to clean drinking water.[30] The lack of sanitation continues to produce serious health consequences throughout the developing countries, including approximately 250 million cases of water-related diseases and at least 5 million to 10 million deaths reported annually.[31] These numbers probably understate the true situation, because public health reporting in the developing world is woefully inadequate. Among the leading water-related killers are diarrhea and schistosomiasis. The main source of these water-related diseases is the drinking of water contaminated with human and animal excrement. Another dreaded disease, cholera, is having a resurgence in the poorest regions in Latin America, Africa, and Asia owing to the lack of basic sanitation.

These and other water-borne diseases today cost global society hundreds of billions of dollars each year. The problem is not related to limitations of medical or environmental science—humankind has more than adequate knowledge to control or eradicate every one of these diseases. The problem is the vicious cycle of poverty in which billions of unfortunate humans are trapped. Barely surviving at well below the poverty line, they live with the daily realities of disease and poor health. Without education or formal employment, they cannot afford even basic sanitary and health services. And in most places they receive very little help from public institutions, which are drastically underfunded and unable to supply even a minimal level of services and also in too many cases are technically inept or politically corrupt, or both.

Thus the water problem in most developing countries is not really a water problem. The environmental issues and the health issues are all consequences of poverty. The world is *not* headed for a water calamity, as Senator Simon fears. But the world *is* headed for a poverty calamity, in the midst of the greatest affluence civilization has ever known. Spread some of that affluence to the world's poor, along with education and basic freedoms, and most of their water problems will be solved.

Yet not all environmental problems related to water can be explained solely by a legacy of poverty. At least one case, the tragically degraded water systems in the former Soviet Union and the countries of eastern Europe, are a legacy of seventy years of misguided and bureaucratic communist rule. In Russia, which has ample water resources, half the population does not have access to safe drinking water, and a quarter of all drinking water is lost because of badly maintained water-supply systems. In Poland, Bulgaria, and Slovakia, over 50 percent of the monitored stream length belongs to their poorest water-quality class. More than 20 percent of wastes from municipal sewerage systems is discharged directly into rivers without treatment. Since even the treatment that exists is inadequate, the water downstream from large cities often resembles raw wastewater. In addition, nonsustainable municipal development practices and outdated production technologies in industry and agriculture during the communist years caused high pollution loads in many water bodies. In Poland almost no river water was drinkable in the late 1980s.[32]

Many water quality problems that have been solved sequentially in Western countries over decades must be confronted simultaneously, and massively, in the eastern European countries if even a modest level of water quality is to be achieved. The problem is made more challenging by the immense political, social, and institutional difficulties endured during the transition to market economies. Even so, the first steps are now being taken in these countries toward development of practical and affordable strategies for water quality planning, wastewater management, and financing.

There is no shortage of technical know-how or political wil[l] countries. The problem is, of course, the high costs of cleanup. Stroi now require real money rather than "five-year plans" generated ~y ~~~ tralized state bureaucracies. The major issue is the lack of funds for investment in water quality. By one estimate, investments of 20–40 percent of annual gross domestic product are required—hardly a realistic possibility.[33] So the critical question is whether sufficient funds will become available from outside sources to bring the water quality up to Western standards in the coming decades.

RISING EXPECTATIONS

It would be difficult to exaggerate the contrast between the water quality standards of the affluent countries and those of most developing countries. In the United States and other industrialized countries, serious water-quality problems are rare and, with steadily increasing water-quality standards, will be even rarer in the future. Throughout the twentieth century, local water-purification systems were widely employed in the United States and generally provided Americans with drinking water of very high quality. Nonetheless, periods of intense industrial growth brought a variety of troublesome water-pollution problems. With the enormous increases of industrial production during and after World War II, pollution of U.S. waterways and lakes greatly increased. By the end of the 1960s, water pollution was almost ubiquitous. A famous example is Lake Erie, whose beaches and fishing facilities were mostly closed down by 1970 and whose tributary river, the Cuyahoga, carried so much industrial and household debris that it actually caught fire in 1969. Another example is the Potomac River, which carried raw sewage through the nation's capital for years and whose estuary was shunned by fall-migrating waterfowl for about fifteen winters.

In 1972 the country's first landmark water-quality legislation was passed by Congress. In the years since passage of the Clean Water Act, the United States has invested over $100 billion in water quality. It is not an exaggeration to say that improvements in water quality have been spectacular. In 1972 only 30–40 percent of assessed waters met water quality goals such as being safe for fishing and swimming, but by 1998 60–70 percent were safe. In 1972 wetland losses were estimated at 460,000 acres per year, whereas at present they are only about one-fourth of that rate. Since 1982, soil erosion from cropland has been reduced by more than a third, substantially reducing sediments, nutrients, and other pollutants that reach streams, lakes, and rivers. In 1972 only 85 million people were served by sewage treatment plants; by now fourteen thousand new facilities have been built and 173 million people are served. Not only do such plants exist, but the entire country now has uniform treatment standards for sewage plants. Annual discharges of conventional industrial pollutants have been reduced by over 100 million pounds, and toxic pollutants by 24 million pounds. And in 1998, 89 percent of the U.S. population was served by community drinking-water systems reporting no health standard violations.[34]

This remarkable progress in water quality, achieved in just a few decades by the United States and by other affluent countries, provides grounds for optimism regarding the possibility of bringing high-quality water supplies to people everywhere. The world's freshwater supply is plentiful, more than adequate to sustain a healthy life for nine billion or more people. In the coming decades technological innovations will potentially increase both water quality and efficient water use, while new institutional arrangements will increasingly reflect the true societal and economic value of water.

But making real progress has other conditions as well. First, planning for future water requirements by local and regional governments needs to become more realistic, framed not in traditional projections of ever higher water "needs" but rather in terms of actually available water. Second, expanded international cooperation programs are essential to promote more equitable development and distribution of water resources. Third and perhaps most important, growing investments of financial and human resources, both public and private, must be dedicated over the coming decades to solving the world's water problems.

Although much has been achieved in the United States in improving water quality and availability, the progress is not sufficient in the context of the constantly rising expectations and priorities of a very affluent society. Many challenges remain for reaching the level of water quality that Americans want and deserve. Most of the country's coastal waters need protecting and restoring. The continuing loss of wetlands, though much lower than in the 1970s and 1980s, must be further slowed. The water quality of lakes, rivers, estuaries, and entire watersheds must be improved so they meet all water-quality goals. Chemical and microbial contaminants in drinking water, some of which pose increasing threats to public health and wildlife habitat, must be further lowered.

Although to most Americans these goals are regarded as imperative, they could understandably be judged as esoteric, remote, and perfectionist in the context of the appalling water conditions that still afflict billions of the world's poor people. Thus, in its totality, the water issue provides a clear illustration of the huge gap between the environmental perceptions and priorities of the rich and of the poor. Water also illustrates how societal expectations of environmental quality continually rise as affluence rises. That is how it should be. Nonetheless, the exaggerated rhetoric of environmental catastrophe that pervades the affluent societies reveals a certain insensitivity to the importance of these rich–poor differences. It would be unfortunate if this rhetoric has the overall effect of deflecting attention from the world's really critical environmental problem—poverty.

THE AIR WE BREATHE

7

Is the air you breathe getting cleaner or dirtier? If you live in Los Angeles, your air is getting cleaner. Once considered the smog capital of the world, the Los Angeles basin is now less polluted than it has been in half a century. Today residents and visitors can often enjoy blue skies and a view of the beautiful San Gabriel Mountains in the distance.

Not so if you live in Mexico City. The air is becoming dirtier, and only rarely can you catch a glimpse of the snowcapped volcanoes that half a century ago provided a spectacular vista. Mexico City, like many other cities of the developing world, has experienced explosive growth and rapid industrialization, which in combination have produced levels of air pollution ranking among the world's worst.

Is the tale of these two cities an isolated case, or can the comparison of their air quality be generalized to other cities of the affluent and developing worlds, with the former improving and the latter worsening? Is air quality really improving in the affluent countries, or is that only a claim made by chambers of commerce? Knowing how industrialization and air pollution historically went hand in hand, should we be optimistic that the poorer countries will clean up their air as they develop or will their air quality continue to deteriorate?

POOR AIR FOR POOR PEOPLE

One tends to think of air pollution as a recent phenomenon accompanying the growth of modern industry, with plumes of smoke belching from tall factory chimneys. But air pollution is as old as fire itself. When humans began to burn wood in their unventilated huts and caves for warmth and cooking, they experienced the unpleasant and unhealthy effects of inhaling soot and smoke. And medieval cities were noted for their polluted air—smoky and putrid from wood fires, dust, animal manure, garbage, sewage, and the wastes of early industries such as smelting and tanning.

Today, preindustrial conditions still exist in many developing countries. The poorest people in these countries live in huts containing a deadly combination of inefficient stoves, poor ventilation, and open fires burning wood, coal, charcoal, dung, or crop residues for cooking and heating—all of which produce a smoky brew of respirable and carcinogenic pollutants. The World Health Organization (WHO) estimates that as many as one billion people are regularly exposed to levels of indoor air pollution up to one hundred times higher than WHO air-quality guidelines.[1] Most of the victims are women and children, who spend much of the day indoors. WHO estimates that in India and sub-Saharan Africa alone, one million children die annually from indoor air pollution, especially from acute respiratory infections. Worldwide, 60 percent of all deaths from disease in children under age fifteen are caused by acute respiratory infections.

One of the most ubiquitous air pollutants is particulate matter, for example, soot from open fires. When inhaled, the very fine particles in soot penetrate deep into the lungs, where they cause not only irritations and infections but possibly, in some cases, cancer. Exposure to particulate matter is not just an urban phenomenon, as often assumed. According to the WHO, nearly three-fifths of the total global exposure to particulate matter occurs in the poorest *rural* areas of developing countries, where indoor air pollution is so severe. The WHO translates this exposure into as many as three million deaths a year worldwide. Even so, air pollution is only one of the health risks faced by the world's poorest people, along with risks from inadequate nutrition, water, health care, and housing. In the least developed countries, these deprivations together cause over a quarter of all deaths.

As countries make the transition from the preindustrial to the early industrial phase of development, environmental quality, especially air quality, inevitably deteriorates. This happens because the traditional small-scale sources of pollution, such as household stoves and fireplaces, remain in use, while growing fleets of automobiles and trucks and new, larger-scale industrial sources, such as manufacturing facilities, refineries, and electricity generators, are added to the mix. In most poor countries, the electricity generators are old and still use antiquated dirty-burning technologies. As a consequence, people in poor countries today often endure air pollution as severe as that experienced two centuries ago in the newly industrializing cities of England and the United States. And the natural resources that provide the copious energy needed for industrialization can themselves be an environmental mixed blessing, as for example the widespread and still growing use of coal in China. On one hand, coal provides household heat for millions of poor Chinese families who previously had no hope of keeping warm in wintertime. On the other, the ubiquitous burning of coal for electricity generation and industrial process-heat throughout China produces some of the most polluted air in the world, indoors and outdoors. During autumn and winter, many residents of Beijing develop respiratory problems from the coal-generated the air pollution, and visitors often develop coughs or other bronchial irritations after only a few days in the region. The levels of total suspended particulates in Beijing's air routinely reach 800 micrograms per cubic meter (mcg/m³) in wintertime, almost ten times greater than the WHO air-quality guidelines.[2] To put such an extreme pollution level into perspective, it is sufficient to point out that long-term exposure to concentrations of particulate matter as low as 10 mcg/m³ has been associated by WHO with a discernable reduction in life expectancy.[3]

How do residents of China react to living with this witches' brew of pollutants? Here's a typical response. University economist Zhenbing, referring to the present state of China's development, said to reporter Mark Hertsgaard, "Economic development is the most important goal for China. It is more important than environment, than human rights, or the other issues the Western media and governments complain about. . . . How much pollution we make, how many trees we cut or dams we build is nobody's business but ours." But he added, "We are used to it. I have lived here for years, so my body has gotten used to this air."[4] Of course Mr. Zhenbing did not claim to enjoy breathing the foul air of Beijing. But pollution is simply a fact of life there—not exactly welcomed but rationalized as an inevitable by-product of the country's economic progress on its long journey to affluence.

The deterioration of air quality is especially acute in Latin America, where millions of cases of chronic respiratory illness are attributable to air pollution. Mexico City is plagued almost year-round by lung-biting, eye-stinging, crop-damaging smog. Although poorly regulated industrial facilities contribute mightily to the area's pollution, the main culprit is the region's 3.5 million cars, mostly older models not equipped with catalytic converters and other pollution controls mandatory for vehicles in the

United States and other affluent countries. Mexico City has grown at a tremendous rate—now at nineteen million inhabitants with overcrowded housing, congested streets, proliferating factories, and inadequate public transportation. But this fact is not in itself unusual among third-world megacities. What is more unusual is the almost unique geography of Mexico City, a sun-baked, seventy-mile-wide highland basin surrounded by a mile-high ring of mountains. This mountain ring effectively prevents the basin's air from being regularly cleansed of pollutants by the action of winds and frequently results in a stagnant air mass over the city. Thermal inversions further impede wind flow and dispersion of pollutants. The result is one of the worst air-pollution problems, arguably the worst, in the world.

Over one million Mexico City residents suffer permanent breathing difficulties, headaches, coughs, and eye irritations. These difficulties are compounded when a shift of winds occasionally carries smoke and particles from outlying industrial plants and agricultural fires directly into Mexico City and causes the already high levels of ozone and suspended particles to reach record levels. During these not infrequent crises, outdoor school activities are canceled, industrial production is drastically curtailed, and half the private automobile population is banned from the streets. In 1998 a rash of wildfires exacerbated Mexico City's already massive air-pollution problems, and several million citizens were taken ill with acute respiratory symptoms and treated at emergency facilities.

Although some environmental protection laws were promulgated in Mexico as early as the 1970s, few emission control restrictions were applied to industry and motor vehicles until 1989, when a clean-air strategy was first devised. After many years implementation of various air-quality control programs appears to be having some effect, especially in keeping ozone levels within bounds, yet air pollution is generally still increasing and remains a source of respiratory health problems in Mexico City.[5]

Why has Mexico not made more progress in cleaning up the air of its capital city? Certainly one factor is the city's topographical bad luck, but more important, the country's recurring economic crises have delayed its modernization generally and its environmental improvement in particular. Mexico City's transportation system is totally inadequate for a city of nineteen million; the existing environmental controls on industrial activity are weak and poorly enforced; the aging vehicle fleet is ill equipped with pollution controls and insufficiently monitored. These problems are at least partly due to the weak support shown by past Mexican governments for setting and enforcing national environmental priorities. But the fundamental cause of Mexico's environmental degradation is the country's enduring legacy of poverty.

THE AIR OF AFFLUENCE

The affluent countries generally enjoy clean air today, but it was not always so. Air pollution long predates the industrial revolution, going back to the Middle Ages, when small-scale coal burning added soot and sulfurous odors to the otherwise dubious pleasures of inner-city life. Coal smoke was not unknown in colonial North America, where blacksmiths used small amounts of local coal in their forges and farmers sold coal chunks they found on their land. Still, it was the industrial revolution in late-eighteenth-century Britain, with its proliferation of coal-burning factories and homes, that elevated air pollution to levels hitherto unknown. The omnipresent smoke, the brown haze, and the pea-souper fog became facts of life in the cities and towns of industrializing Britain. Miners and factory workers and their families were deeply distressed by the ubiquitous environmental degradation and were powerless to escape it, yet many accepted pollution as an unavoidable element of the industrialization that provided their livelihoods.

Author Charles Dickens did not accept it. In his 1854 novel, *Hard Times*, Dickens paints a scathing picture of the Victorian industrial society.[6] Here is his description of the fictional city of Coketown, based on actual industrial cities of Dickens's time:

> It was a town of red brick, or of brick that would have been red if the smoke and ashes had allowed it; but, as matters stood it was a town of unnatural red and black like the painted face of a savage. It was a town of machinery and tall chimneys, out of which interminable serpents of smoke trailed themselves for ever and ever, and never got uncoiled. It had a black canal in it, and a river that ran purple with ill-smelling dye, and vast piles of building full of windows where there was a rattling and trembling all day long, and where the piston of the steam-engine worked monotonously up and down, like the head of an elephant in a state of melancholy madness.

Coketown industrialist Josiah Bounderby cynically explains that this was pollution with a purpose: "First of all, you see our smoke. That's meat and drink to us. It's the healthiest thing in the world in all respects, and particularly for the lungs." But the smoke-laden air was anything but healthy when a pea souper turned into a deadly "killer fog"—a not infrequent occurrence in the Coketowns of nineteenth century Britain.

In the United States industrialization accelerated in the late nineteenth and early twentieth centuries. Coal overtook wood as the leading energy source and soon became the principal fuel of the country's rapidly growing economy. Belching soot from coal fires, the tall smokestacks became a major symbol of the new industrial age. And the killer fogs came to the United States as well. On October 29, 1948, a dense, acrid fog descended on the small industrial town of Donora, Pennsylvania, sending six thousand residents to the hospital with breathing difficulties and killing seventeen— the first known American deaths from air pollution. The deadly ingredient of this killer fog was sulfur dioxide emitted by the town's zinc processing plant and entrapped in the atmosphere by a seasonal temperature inversion. Earlier in the year, six hundred Londoners died when a deadly fog blanketed the city. London was to experience even worse episodes; in 1956 a thousand people died in a single episode, and in 1962 seven hundred fifty died. Although the fraction of these deaths that were directly caused by the killer fogs is not known, nonetheless these episodes had a cumulative impact that helped to catalyze clean-air movements throughout the industrial world.

Large-scale coal burning had caused a dramatic increase in emissions of air pollutants, especially sulfur dioxide (SO_2), nitrogen dioxide (NO_2), and particulate matter (PM). Sulfur dioxide is one of the major pollutant emissions from coal combustion and is thought to have been the principal killer in the episodes of Donora, London, and elsewhere. Sulfur dioxide emissions originate not from coal's carbonaceous matter itself but from sulfur impurities in the coal (reaching as high as 6 percent in some soft coals). Reflecting the twentieth century's growth in industrial coal combustion, emissions of SO_2 in the United States tripled between 1900 and 1970, increasing from ten million to thirty million tons annually.[7]

Another major pollutant, NOx, a mixture of oxides of nitrogen, arises not from the fuel being burned (e.g., coal or gasoline) but rather from nitrogen, a natural constituent of air. NOx is a by-product of all combustion taking place in air, whether in automobile engines or stationary sources such as industrial boilers or gas-fired kitchen ranges. Its formation is unavoidable wherever combustion takes place at high enough temperatures for oxygen and nitrogen to react chemically. In 1900 just over two million tons of NOx were emitted into the atmosphere in the United States. Most of this amount came from wood and coal combustion, since at the time there were only eight thousand automobiles in the entire country. But

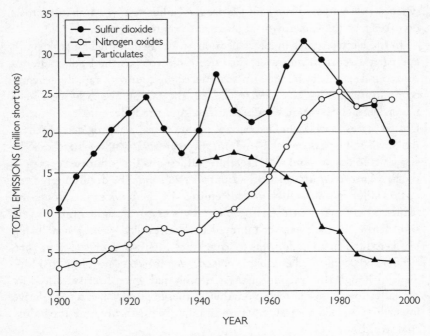

Figure 17. Trends in U.S. air-pollutant emissions (1900–1995). Data are from U.S. Environmental Protection Agency, *National Air Pollution Emission Trends, 1900–1996*, report EPA-454/R-97-011 (Washington, DC: EPA, December 1997).

today the U.S. vehicle population has grown to 170 million, and this enormous fleet of vehicles is responsible for over half of the current twenty-five million tons of NOx emitted annually.

The public-health significance of NOx comes from two facts: first, NOx is an irritant to the throat and lungs even from short, low-level exposures, while long-term, high-level exposures have been shown to produce emphysema-like effects and reduced resistance to bacterial and viral infections of the lung.[8] Second, NOx is one of the principal chemical precursors of so-called photochemical smog found in warm, sunny areas such as Los Angeles and Mexico City. The first recognized eye-smarting episodes of smog in Los Angeles occurred in the summer of 1943, but it was not until 1952 that California scientists discovered the nature and causes of photochemical smog.[9] Simply put, photochemical smog is formed when a mix of nitrogen oxides (NOx) and volatile organic compounds chemically react in the presence of ultraviolet radiation from the sun.

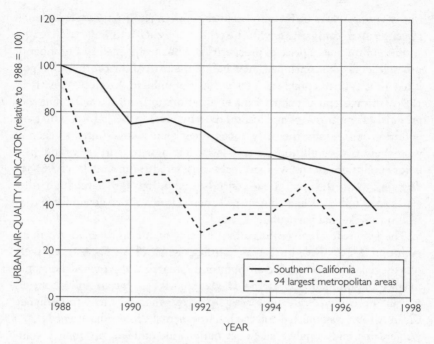

Figure 18.   Improvements in urban air quality (1988–1997). The plotted air-quality indicator is the number of days annually that violated EPA air-pollution criteria (the Pollutant Standards Index [PSI]) in the ninety-four largest U.S. metropolitan areas and southern California, as a percentage of 1988 values. Data are taken from Environmental Protection Agency, National Air Quality and Emission Trends Report, 1997, report EPA-454/R-98-016 (Washington, DC: EPA, December 1998).

In the early years of industrialization, Americans generally tolerated air pollution as an inevitable side effect of the new urban life. Many city dwellers were recent immigrants who had known extreme poverty in other lands, and they lauded the new prosperity created by industrial jobs. As a youngster growing up in an Ohio steel town in the depression years of the 1930s, I recall that people actually welcomed the ubiquitous gray cloud of coal smoke hanging over the steel mills. Even though we choked on the soot and our Sunday clothes became soiled instantly, the smoke cloud meant that the mills were working and our fathers had jobs and we had homes to live in with food on the table. The rivers were so terribly polluted with refuse that the water was almost hidden, but no matter; only a short distance away you could find forests, brooks, and birds. To me, as to my counterparts in Britain a century earlier, pollution seemed a small price to pay in comparison with the economic benefits of industrialization.

But pollution as a price of prosperity did not seem small to a number of urban leaders who concluded, even before the turn of the century, that polluted air was unacceptable as a by-product of industrialization. As early as 1881, Chicago and Cincinnati made an effort to control smoke and soot primarily from furnaces and locomotives by passing the country's first air-pollution statutes. By the early 1900s a few county governments began to pass their own pollution-control laws. Yet a half-century would pass before pollution control was seriously considered at the state level. In 1952 Oregon became the first state to control air pollution legislatively, and other states soon followed, enacting air pollution statutes aimed generally at controlling smoke and particulate matter.[10]

The U.S. federal government first became significantly involved in air pollution in 1963 with the passage of the original Clean Air Act, relatively weak legislation that funded air-pollution research and provided assistance to state and interstate programs. The landmark year for air-pollution control in the United States was actually 1970, when a much strengthened Clean Air Act was enacted and the Environmental Protection Agency (EPA) was created and empowered to set nationwide ambient air-quality standards as well as pollutant emissions standards for cars, trucks, and buses. The 1970 act set the three-pronged formula for air-pollution regulation that is still essentially in place today:

· Consider protection of human health as the primary goal of standards

· Require the use of the best available control technologies

· Write compliance deadlines into law

Amendments to the Clean Air Act passed in 1990 further strengthened EPA's enforcement prerogatives and for the first time addressed a number of pollutant substances considered toxic.

Aided by impressive technological advances in the private sector, the 1970 Clean Air Act and subsequent amendments have brought about dramatic reductions in pollutant emissions and improvements in air quality throughout the United States. According to EPA data, total U.S. emissions of the six principal pollutants[11] have declined each year since 1970, and in 1999 emissions of these pollutants were 31 percent below 1970 levels. This remarkable improvement in air quality occurred during a period when many relevant factors were working in the opposite direction—the U.S. population increased 33 percent, vehicle miles traveled increased 140 percent, gross domestic product increased 147 percent, and the use of coal for generating electricity increased almost threefold.[12] In the 1990s every year in fact showed better air quality than any year during the 1980s. This steady improvement trend occurred despite weather conditions in the 1990s that were generally conducive to higher pollution levels.

U.S. emissions of sulfur oxides (SO_2), the principal chemical killer of Donora, peaked in 1972 at thirty-two million tons and since then have been steadily declining. From 1980 to 1999 SO_2 emissions decreased 28 percent. Remarkably, in 1999 the SO_2 emissions were down to nineteen million tons, which is about the emissions level of 1915.[13]

Reflecting mostly the growth in automobile use, nitrogen oxide (NOx) emissions increased almost tenfold from 1900 to 1980. Then, responding to the federal emissions standards of the 1970s, total NOx emissions leveled off around 1980 and have remained essentially constant since then. The fact that there has not been an actual decrease in total NOx emissions is due largely to the growing contribution from off-highway diesel vehicles (mostly construction vehicles), which are not yet regulated. During this same twenty-year period, however, there has been an actual decrease of 25 percent in the *concentrations* of nitrogen dioxide (NO_2) monitored in or near urban centers, and all areas of the country that once violated the national air-quality standard for NO_2 now meet that standard.

As pointed out earlier in this chapter, particulate matter (e.g., soot) is among the world's most dangerous air pollutants, reaching concentration levels as high as eight hundred micrograms per cubic meter in Beijing, and possibly associated with millions of deaths worldwide, mostly in the developing countries. In the United States the reduction of particulate air pollution has been one of the great success stories of pollution control efforts. Although quantitative data are not available for the period before 1940, particulate levels are known to have been extremely high in the smoky years before controls, probably similar to the levels experienced today in cities such as Beijing. What is known is that particulate emission levels peaked around 1950, steadily declined until the mid-1980s and have remained relatively stable since then. In terms of total emissions nationwide, particulate matter (PM-10)[14] has decreased from about seventeen million tons in 1950 to under four million tons today.[15]

The latest amendments (1990) to the Clean Air Act require the EPA to regulate for the first time a large number of low-concentration air pollutants classified as toxic to humans. EPA defines toxic air pollutants as those that may cause cancer or other serious health effects in people exposed to them at high concentration. The list of low-level toxics contains

188 substances, including, for example, benzene, found in gasoline; per-chloroethylene, emitted from some dry-cleaning facilities; and methylene chloride, used as a solvent by some industries. Although the database on these substances is still tentative, the EPA estimates that total toxic emissions decreased about 23 percent between 1990 and 1996.

These remarkable improvements in air quality came about because the American people strongly supported, and continue to support, environmental regulations at both state and national levels. And when the levels of smoke and smog were actually reduced in metropolitan areas, most of us delighted in the clear blue skies and distant vistas. But these improvements came at a price—a high price. More than $500 billion were directly spent in the United States from 1970 to 1990 on complying with the Clean Air Act. Esthetic pleasure in itself would not seem to justify pollution abatement at such high cost. Indeed, it was primarily the benefits of reduced air pollution on the nation's public health, and to a lesser extent the benefits to agriculture and ecosystems, that provided the political justification for a government regulatory mandate entailing such massive expenditures. But, you may well ask, how do we know that these benefits, undeniable as they are, were worth $500 billion? Maybe they were worth much more, say, $5 trillion? Or maybe they were worth only $5 billion? How can we determine how much the benefits were in fact worth?

If you asked this question, you have just walked into a political thicket, full of socioeconomic brambles, that goes by the name of *cost-benefit analysis.* Superficially it sounds simple enough: cost-benefit analysis is a policy tool that attempts to *measure* the changes in societal well-being resulting from the imposition of government regulations. But actually making such measurements is quite a challenge: for example, what metric should one use for determining the monetary value of chronic bronchitis reduction brought about by reducing particulate matter in air? Or of lowering the incidence of lead-related IQ impairment in children by phasing out leaded gasoline? Such evaluations are extremely difficult, both in principle and in practice. Advocates of cost-benefit analysis argue that it can provide a disciplining element that forces all sides in a policy debate to consider more carefully, in a world of limited financial resources, what is gained and what is given up when making a policy decision.[16] And in fact, two U.S. presidents of opposite parties, Ronald Reagan and Bill Clinton, issued executive orders that *require* all government agencies dealing with significant regulatory expenditures to use cost-benefit analysis to help justify the costs of implementing their regulations.

From its own cost-benefit study, the EPA claims that the monetized benefits brought about by the Clean Air Act over the period 1970 to 1990 "substantially exceeded" the $500 billion costs and were worth about $14 trillion.[17] The EPA arrived at this estimate of benefits by calculating the dollar value of all the harm they believe would have been done by the higher levels of pollution in the absence of the Clean Air Act, such as premature deaths, chronic bronchitis cases, and lower agricultural output. About two-thirds of the benefits were ascribed to lowered mortality from the reduction of particulate matter in air.

The EPA analysis has been attacked from two different quarters, and for very different reasons. The first criticism holds that the EPA simply got its numbers wrong, overestimating the Clean Air Act's benefits and underestimating its costs. On the benefits side, one study concludes that EPA attaches a "value" to a year's extension of human life about one hundred times greater than the corresponding figures used by the Federal Aviation Administration, the Consumer Product Safety Commission, or the National Highway Traffic Safety Administration.[18] This implies a large overestimate of benefits, since most calculated benefits of pollution reduction relate to extending life span.

There may also be additional *costs* of the Clean Air Act arising from losses in the nation's economic productivity caused by environmental and occupational health regulations. Not surprisingly, specific studies disagree widely on the magnitude of these losses. One well-known study concludes that environmental regulations from 1973 to 1985 reduced the nation's GNP by only about 2.6 percent.[19] Another finds a much larger loss, an 11 percent drop in U.S. manufacturing productivity from 1974 to 1986 caused by regulations.[20] In contrast, a study of petroleum refineries in Southern California finds that their productivity actually *increased* during 1987–1992, a period of heavy environmental regulations.

And what about the *total* cost of the Clean Air Act? One study estimates a cost between $1.5 trillion and $3.5 trillion—three to seven times higher than the $0.5 trillion direct cost stated by the EPA.[21] The same study challenges the EPA's estimates of actual life extension brought about by air pollution reduction and concludes that the total benefit may be as low as $1 billion to $5 billion—much lower than EPA's $9.1 trillion figure.[22] These numbers obviously represent the viewpoint that the costs of the Clean Air Act actually exceed its benefits—the opposite of the government's position.

Such grossly inconsistent conclusions based on economic cost-benefit analyses illustrate why such studies become mired in controversy. Not only are they plagued by uncertainties in data and scientific method, but they can also be colored by political objectives. It may never be possible to place an economic valuation on human life that will be widely acceptable, although such valuations are routinely used actuarially by the insurance and public health industries.[23] An additional complication is that the costs and benefits tend not to be evenly distributed among the affected parties. And it may never be possible to project accurately the *future* costs and benefits of alternative regulatory programs.

The other criticism is of a different, more fundamental kind. It argues that cost-benefit analysis of the type carried out by the EPA should not have been done (or required) in the first place. According to this view, it is totally inappropriate to attempt to quantify environmental benefits and values because they are inherently unquantifiable. Environmental ethicist Mark Sagoff puts it this way: "Surely environmental questions—the protection of wilderness, habitats, water, land, and air as well as policy toward environmental safety and health—involve moral and aesthetic principles and not just economic ones."[24]

I fully subscribe to the notion that moral and esthetic principles, in addition to economic principles, are involved in making environmental judgments. Which leads me back to the core issue of this book. I argue repeatedly in these pages that most citizens of affluent, democratic countries are environmentalists at heart. Although most of us do not usually think in philosophical terms about the environment, the fact is that we value clean air, clean water, and beautiful surroundings, and we are willing and able to pay whatever it costs to attain and maintain a healthy environment. The passage of the Clean Air Act in the first place is testimony to the fact that, for an affluent society, sparkling blue skies and breathtaking vistas may in the end be sufficient justification for such measures.

THE ACID RAIN PROBLEM

"Acid rain" arises primarily when sulfur dioxide (SO_2), emitted from fossil fuel combustion, and oxides of nitrogen (NOx), emitted from vehicle exhausts, react with water and oxygen in air to form acidic compounds. Once released into the atmosphere, these can be converted chemically into secondary pollutants including water-soluble substances such as sulfuric and nitric acids. The resulting acidic compounds can be transported hundreds of miles by prevailing winds and deposited as wet or dry acidic substances on fields, forests, and bodies of water.

During the 1970s a number of environmental scientists became concerned that acid rain could have serious biological consequences, including extensive damage to forests, crops, fish, buildings, and human health.[25] Researchers found that the acidity of precipitation was higher than normal in a number of locations in the United States, Canada, Scandinavia, and China. Ecological damage was noted in Scandinavian lakes, including increases in trace toxic metals, deaths of trout populations, and changes in microbial activities. Similar effects were noted in Canadian and U.S. Adirondack lakes.[26] Forest damage has also been documented in the United States, and acid rain is believed to be among the causes.[27]

As mentioned previously, emissions of the chemical precursors to acid rain (oxides of sulfur and nitrogen) constantly increased in the United States during the more than half-century of industrial expansion. In the 1970s acid rain became the subject of intense public attention and was indicted by the media as one of the world's most serious environmental problems. Following the advent of national air-quality regulations in the 1970s and 1980s, as we have seen, sulfur emissions decreased and nitrogen oxide emissions stabilized. And from 1995 through 1998, deposition of sulfates in precipitation exhibited dramatic reductions over a large part of the eastern United States.[28] Just how serious was the acid rain problem, and what impacts did the pollutant reductions brought about by government clean-air policy have on the problem and its prognosis for the future?

The proposition that acid rain constitutes a serious environmental threat was investigated by a decade-long government-coordinated research effort conducted during the 1980s.[29] The study's research contributed significantly to an understanding of acid rain, but differences of view emerged both about the study's scientific results and about how they were presented. For example, ecologist Gene Likens takes issue with reviews implying that the study found acid rain impacts to be less destructive than earlier predicted, and he asserts that the acid rain problem is actually greater than earlier suggested.[30] Citing studies of fir and red spruce trees that show damage in which both acid rain and natural stresses may be involved, Likens argues that the real and potential effects of air pollutants on natural ecosystems are more complicated than generally appreciated. Though he acknowledges the benefits of the 1990 amendments to the Clean Air Act in reducing acid deposition, he argues that the reductions will not be great enough to accomplish the goal of preventing long-term harmful effects on essential ecosystem properties.[31]

The study's original director, J. L. Kulp, offers a more optimistic prognosis:

> The effects of present levels of acid rain were found to range from net positive (e.g., on crops) to modest negative (e.g., surface water). There is no evidence that any significant worsening of these effects will occur over the next half century even if the present levels of pollution were to continue. Actually, current regulations will produce a steadily decreasing acidity of rain in the USA, Europe, and Japan. Finally, new technology for control has emerged so that in the future new coal-burning plants can operate with negligible levels of acid-forming emissions at no greater cost than plants using the older technology.[32]

Disagreements among scientists about the ecological and economic consequences of acid rain will undoubtedly continue, because there is much at stake, politically and economically, and much still to be learned. But the fact is that the United States and other affluent countries are pursuing regulatory policies and costly control programs that have already reduced the severity of the acid rain problem. Such progress cannot be reported from the poor countries, because they cannot afford the costs of equivalent environmental programs.

ARE WE EXPORTING OUR POLLUTION?

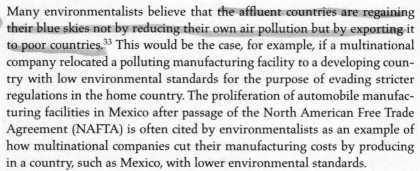

Many environmentalists believe that the affluent countries are regaining their blue skies not by reducing their own air pollution but by exporting it to poor countries.[33] This would be the case, for example, if a multinational company relocated a polluting manufacturing facility to a developing country with low environmental standards for the purpose of evading stricter regulations in the home country. The proliferation of automobile manufacturing facilities in Mexico after passage of the North American Free Trade Agreement (NAFTA) is often cited by environmentalists as an example of how multinational companies cut their manufacturing costs by producing in a country, such as Mexico, with lower environmental standards.

The facts tell a different story. In the case of Mexico, the pertinent environmental regulations are actually roughly equal to those in the United States. In the past, Mexico's enforcement has been more lax, but that situation is changing. With regard to the companies' intent, a U.S. interagency task force had this to say: "U.S. firms, particularly the larger multinational firms most likely to undertake large process industry investments, often hold subsidiaries to a worldwide standard usually at least as high as standards with which they must comply in the U.S."[34]

In any case, environmental compliance is rarely a major cost factor in operating a manufacturing facility. In the United States, for example, industry's pollution abatement costs are, on average, only 0.6 percent of revenue, rising to between 1.5 and 2 percent for the most polluting industries.[35] And in the countries of the Organization for Economic Cooperation and Development (OECD), "direct environmental costs are believed to account for 1–5 percent of production costs."[36] Whatever small savings might accrue to the companies from *differences* in environmental compliance costs at different sites are dwarfed by the overall investment costs for new facilities. A 1990 study of this issue found no evidence that global trade patterns are affected to a significant degree by differential environmental standards.[37] A more recent study concludes that "to the extent that the developed countries are exporting their dirty industries, they seem to be exporting them to each other, not to the less developed economies."[38] As further evidence of the economics of compliance, the study also concludes that there is "no overall tendency for plants with superior environmental performance to be less profitable."[39] And another study puts it this way: "Multinational firms cannot escape their environmental obligations by moving polluting plants offshore. . . . Market forces nowadays reward good environmental performance rather than cost savings at any price. . . . When consumers care, producers care."[40]

The main motivation for the huge investments in production facilities located in foreign countries is not to evade environmental regulations but rather to improve production efficiency by reducing energy and resource demands so as to increase the product's worldwide competitiveness. These investments are not biased toward polluting industries but rather toward labor-intensive industries that are *less* polluting on average.[41] In pursuing these goals, foreign automotive companies brought state-of-the-art manufacturing technology to Mexico's automotive sector and provided employment opportunities for thousands of skilled workers. Without the economic growth that such opportunities make possible, Mexico could never adequately address its environmental problems.

Yet there are cases, of course widely publicized, where local corporate managers in developing countries have taken advantage of lax environmental enforcement and allowed their plants to pollute rivers, skies, and fields, even though the cost savings were probably minimal. Although such behavior may not be altogether unexpected given the intensely competitive markets in which many firms operate, it is inexcusable both on moral grounds and because it constitutes poor, myopic business practice.

When morality and economics appear to collide in such situations, we should temper our outrage with the knowledge that, in this transparent world, the forces of morality and economics are increasingly working in the same direction. Although wages may still be lower than in the affluent countries, manufacturing facilities in developing countries provide jobs and security for millions of families, in many cases raising their living standards far above the subsistence levels that have long been the local norm. Earning food for the table is a first step in the long journey out of poverty, and environmental priorities are but a few steps down that road.

Still, neither economic nor environmental progress comes overnight. It is worth remembering that, when the industrial countries got serious about setting and enforcing environmental standards after World War II, it took four decades to reach the present environmental situation, which is much improved but still far from perfect. It would be unrealistic to expect that, even with international employers following best practices, environmental standards throughout the developing world will quickly become comparable to those of the affluent countries. The level of poverty must be lowered considerably, and the level of personal freedom raised considerably, before people in the developing world can be expected to acquire the political will and economic means to demand the level of environmental standards that we in the affluent world now take for granted.

POSTSCRIPT

In December 1997 the *San Francisco Chronicle* ran a story titled "Skies Blue Again in L.A.—Cleanest Air in 50 Years."[42] It began with the engaging line "The last time Los Angeles Basin air was as clean as it is today, fedoras were required headgear, Truman was president and Bogart and Bacall were still making movies together." One can justifiably feel nostalgic about those Bogart–Bacall films but nostalgia is not in order for the smoggy skies that plagued Los Angeles for the ensuing fifty years. The article notes that "in 1970, there were 148 Stage One ozone alerts in Los Angeles. Stage One alerts are usually associated with air that is approximately the color of river mud. The elderly, young and ill are advised to stay indoors. In 1997, there was only one such alert." And a Los Angeles resident noted, "The mountains used to be completely obscured. Ten or 15 years ago it was a rare day that you could even see them. Now they seem almost magnified. They're gorgeous—just completely gorgeous."

The article's euphoric tone notwithstanding, most scientists would not claim victory in the battle against polluted air, in Los Angeles or anywhere else. Still, great progress has surely been made in recent decades. This progress has been spurred by good environmental science, technological innovation in industry, and the public's increasing valuation of air quality's benefits, all of which in turn have sanctioned a diverse array of ever tighter and more costly government standards and regulations. Indeed, efforts to improve air quality have been undertaken at the behest of affluent and democratic societies throughout the world. The citizenry not only have given high political priority to improving air quality but have also been able to marshal substantial public and private resources for investments in clean air—in the United States at least a trillion dollars in the last three decades. These investments are now paying off in terms of improved health, visibility, and quality of life.

Yet air quality is clearly a work in progress. Much has been accomplished, but how much more needs to be done? When is our air "clean enough"? Science cannot answer these questions because these are not scientific questions, though science can play an important role in providing knowledge relevant to answering them. These are mainly questions of societal expectations and economics. In the affluent societies, probably the air will never be "clean enough" because we keep raising the bar of our social and environmental expectations. As affluence continues to rise in the United States and other industrial countries, popular demand for continuing improvements in air quality will only increase.

In contrast, developing countries, with more pressing social and economic priorities, have not yet made comparable investments in air quality. The dreadful state of the air in so many poor countries testifies to this. When they are further along their development paths and their citizens more widely enjoy the basic freedoms, these countries will almost certainly opt for clean air—partly because the requisite technologies will be within reach economically but, more important, because their people will demand clean air.

8

FOSSIL FUELS—

CULPRIT OR GENIE?

A recent TV commercial tells us "it took hundreds of centuries to create the oil resource and only 150 years to deplete it." The familiar voice in the commercial is that of a former TV news anchor. The sponsor, a large agribusiness, produces corn-derived ethanol, a synthetic gasoline substitute or blend. It's understandable that synthetic fuel producers—the beneficiaries of generous government subsidies—have an incentive to promote the idea of a worldwide oil shortage, but do the facts support this claim? Are we actually "pumping the well dry"?

At first glance it would seem so. For one thing, the world's energy use continues to grow each year, and the rate of increase is especially high in the developing countries. To quote the same agribusiness, "Like it or not, fossil fuels are finite." So if there's only so much oil in the ground, isn't it bound to run out? Not necessarily. Of course fossil fuel supplies are finite—they're certainly not infinite! But finiteness is not the issue. Availability and affordability are the real issues. In this chapter we'll examine the family of fossil energy resources and ask whether their availability is likely to become limited—hence costlier—in any significant way in the foreseeable future.

Today many people believe that fossil fuels, especially oil, are becoming scarce or even running out. This is hardly surprising, not only because of incessant media pronouncements to that effect but also because energy consumers have directly experienced at least two energy "crises" in recent decades. Yet if we look at the historical evidence, we find a quite different picture of fossil fuels—one of growing abundance rather than growing scarcity. Technological advances in mining and producing fuels, together with competition among suppliers, continuously increased the availability of fossil energy resources after World War II. As a result, the price of fossil fuels to the consumer steadily fell. In fact, prior to the 1970s, energy had become so cheap that consumers rarely considered energy costs as a factor when purchasing a home, an automobile, or an appliance.

Then came the first energy crisis. The October 1973 war in the Middle East triggered two waves of energy price increases that catapulted the subject of energy out of academic obscurity into the media spotlight. Between October 1973 and January 1974 the price of oil on the world market quadrupled. In the United States long lines of waiting consumers at gasoline stations became commonplace across the country. The alarmist and often inaccurate media coverage of this situation understandably led people to believe that the world was running low on oil resources.

But there never was a real shortage of oil. The 1973–1974 energy crisis was mostly a show of political power by some Middle Eastern members of the oil producers' cartel the Organization of Petroleum Exporting Countries (OPEC), which at that time controlled much of the world's oil production. By imposing a slowdown in their oil production and taking advantage of weaknesses in the global oil-distribution system, OPEC was able to create artificial gasoline shortages all over the world and thereby to increase prices and profits. The producers' victory, however, was short-lived. After only a few years, new, non-OPEC production sources came on line, depriving OPEC of its control over the global oil market and allowing competition to be restored.[1] Oil prices soon fell to historic lows.

The year 2001 brought a new version of energy crisis, this time an electricity shortage that struck California just at the peak of a prosperous decade. Again, the crisis came not from a resource shortage, nor was it caused by foreign oil producers. It was the result of a bungled attempt to bring more competition into the state's electricity system through deregulation. The legislature's design for deregulation—created under intense lobbying by utilities, environmental groups, big users of electricity, and regulators—was deeply flawed. Most serious was the provision that put a ceiling on *retail* electricity prices but allowed *wholesale* prices to be set freely by the market. Because of the state's late-1990s economic boom, consumer demand for electricity was growing rapidly. There was no shortage of fuels for producing electricity, but the state had too few power plants to accommodate the mounting electricity demand. And with retail prices that couldn't be raised, consumers had no incentive to reduce electricity use. As with any market commodity, a surefire prescription for soaring prices is consumer demand that can't be met. So the utilities were forced to purchase electricity at rapidly rising wholesale prices. An additional strain on the system was a spike in the price of natural gas, which the utilities also could not pass on to electricity consumers. These unprecedented expenses drained the coffers of the state's two largest utilities and drove one of them into bankruptcy. So the system essentially broke down, and blackouts routinely occurred during hot spells and other periods of high electricity use. Still the state avoided raising retail prices, instead buying billions of dollars of power with money from current and future tax revenues and making long-term purchase contracts at the height of the crisis that, unless overturned, will result in high electricity rates in California for a decade.

The 1973–1974 oil jolt to the world economy was eventually understood in terms of Middle East politics,[2] and the 2001 electricity crisis was quickly diagnosed as a failure of California regulatory policy. Both situations were exacerbated by government mishandling of energy markets. Although neither crisis was directly related to global resource abundance, both had the effect of rekindling fears about running out of energy. These fears remain today even though the technical and economic facts about energy resources and technologies portray a much more optimistic picture. In fact vast supplies of energy resources are available, enough to last civilization for thousands of years.

But apart from concerns about resource scarcity, environmentalists also worry about the environmental impacts of rising energy use. Most believe that affluent consumers are very wasteful in their use of energy and that this wastefulness promotes not only resource depletion but also pollution and other environmental damage. And they worry even more that the world's poor people will inevitably emulate the wasteful energy habits of the rich as they become affluent, and will further exacerbate the world's environmental problems.

The evidence indicates that these concerns are overblown. This chapter reviews the family of energy resources from the perspectives of the poor, the rich, and the environment. It also points out that, in the affluent societies, people overwhelmingly rate environmental quality as important to their lives and their lifestyles are becoming kinder overall to the environment as well as more efficient (less wasteful) in the use of energy resources. Although people moving out of poverty certainly do emulate the lifestyles of the affluent, in this century they will be able to take advantage of an ever broadening array of resource-efficient and environmentally friendly energy choices.

WOOD

Let's start with humanity's oldest energy resource—wood. Although wood is not a fossil fuel, it is discussed as a fossil resource in this chapter largely for chronological reasons. Today the world's forests are valuable not as fuel resources but mostly as ecological resources (tropical rain forests) and sources of building materials (wood farms).

From time immemorial, wood was used as the major fuel for heating, cooking, and working of metals, ceramics, and glass. And still today, wood remains the only energy source for millions of the world's poorest people. In this situation as in many others, poverty imposes serious and unavoidable environmental risks. One example is the enormous health risk caused by the use of open-pit wood fires for cooking and heating in the unventilated stoves found in dwellings of the poorest families worldwide. Wood smoke, known to contain high levels of respirable carcinogens, is copiously inhaled by the women and children who spend so much time indoors near these unventilated wood fires.[3]

Five centuries ago wood was the major source of energy in the settlement of the New World. Yet, contrary to common belief, the early European settlers did not find virgin forests undisturbed by human activities. In fact, the native American populations had been clearing forests for energy and construction long before 1492,[4] and these uses continued throughout the colonial period. Later, as the young United States expanded, domestic forests were destroyed at a rapid rate, partly for the wood supply but mostly to clear new lands for the rapidly expanding agriculture. Three hundred years after the arrival of the Europeans, U.S. forests had been reduced from 40 percent to 30 percent of the country's land area.[5]

During the period of massive deforestation, concern was widely expressed about an impending "national famine of wood."[6] This wood famine never materialized. What saved the U.S. forests was, first, the development of enlightened government conservation policies, including the national forest system; second, the emergence of steel and concrete as superior construction materials; third, the switch from wood to fossil fuels for the nation's growing energy needs; and probably most important, the growth of affluence, which made these things possible.

As late as the mid-nineteenth century, wood provided about 90 percent of the nation's energy output. With the advent of industrialization, fuel-wood began to be displaced by coal. Coal was then coming into wide industrial use and ultimately became the major fuel of the industrial revolution. The use of wood for energy in the United States peaked around 1870 and then steadily declined.[7] By 1920 wood provided only about 10 percent of U.S. energy output. In the United States the era of fuelwood was over.

In contrast, *nonfuel* industrial uses of wood increased in the United States throughout the nineteenth century. At least a quarter of the wood was used for railroad ties and bridges as the railroad system expanded.[8] As the twentieth century approached, however, concrete and steel were replacing most industrial uses of wood, making possible huge advances in architecture and building construction—including the tall skyscrapers for which American cities became world famous. Around 1920 the era of large-scale forest clearing in the United States finally came to an end. Today, wood—grown from dedicated and sustainable tracts—is used mostly for paper and for construction of small buildings and residences.

Agriculture also contributed to saving the American forests. As new fossil fuel–powered agricultural machines were introduced early in the twentieth century, farmers were able to grow crops more efficiently, so they needed less land for a given output. The case of Vermont is illustrative. In the 1700s Vermont was almost totally covered by forest, yet by 1850 so much clearing had taken place for agricultural use that forest cover had dropped to 35 percent. People feared that Vermont would become a wasteland.[9] Today, however, Vermont's forest cover has been restored so that the state now contains 77 percent forest.[10] Once again Vermont is a land of beautiful forests.

In the United States as a whole, over 300 million acres of forest were lost between 1600 and 1920. The forest area began to stabilize around the turn of the twentieth century and since 1920 has actually been expanding. At present, the total acreage of U.S. forests is 737 million acres, almost three-quarters as large as it was in 1600.[11] Many forests that had been totally destroyed have been restored and added to the U.S. national forest system, providing human recreation, wildlife habitat, and wilderness. The National Wilderness Preservation System grew from 9 million acres in 1964 to 104 million in 1994.[12]

Roughly two-thirds of U.S. forest area is classified as timberland— forests capable of and not excluded from commercial timber production. Since the 1950s, timber growth has consistently exceeded harvest. A wide variety of softwood and hardwood species is thriving across the country.[13] At the same time, the commercial supply of both hardwoods and softwoods is increasing,[14] and the United States continues to be the world's major industrial wood producer, supplying roughly 25 percent of the world's total.[15] The wood supply in the United States will be available indefinitely because industry and government continue to invest in efficient forest management techniques and technologies. Forest expert Douglas W. MacCleery concludes that American forests today are "in significantly better condition than they were a century ago."[16] Similar examples of forest recovery and growth can be cited for many European countries. There is little question that in the United States and other affluent countries, wood resources have reached a healthy and sustainable state.

Nonetheless, there is no general agreement on how these abundant wood resources should be used. In the United States the federal government faces a difficult challenge to balance competing and possibly incompatible demands on federal timberlands coming from groups with very different agendas—conservationists, recreational users, and commercial interests. The situation is reflected in the current political uncertainty as to the actual mission of the U.S. Forest Service. In 1970 the agency was given a legislative mandate to protect "the multiple use and sustained yield of the products and services obtained on Forest Service land." These uses included recreation, range, timber, watershed, wildlife, fish, and wilderness. But more recently (1999) a Department of Agriculture committee concluded that "ecological sustainability should be the guiding star for the stewardship of the national forests." This "guiding star" would presumably be incompatible with logging, even sustainable logging. Some conservation groups now urge that logging be banned entirely from U.S. national forests and that the forests be devoted mostly to recreational services.[17] Forest expert Roger Sedjo comments: "Giving preeminence to the preservation of biodiversity directly contradicts the Forest Service's legislative mandate to manage the land for multiple outputs including mining, grazing, logging, or other commercial activities. Such a shift may be warranted, but without a new legislative mandate it is unclear which overriding management goal the Forest Service should serve."[18]

The bottom line is this: in the *affluent* countries the forest resources will remain more than satisfactory in terms of total forest area and resource sustainability. Unresolved, however, is the growing conflict between those who regard sustainable commercial uses of public-domain forests as economically and environmentally sound and those who believe that ecological preservation should take precedence over any other uses, sustainable or not.

In the *developing* countries the situation is almost totally opposite. The era of deforestation is by no means over. The developing world lost almost 10 percent of its natural forested area during the period from 1980 to 1995, a time when the developed countries actually added about 1 percent to their forested area.[19] In Brazil almost 15 percent of the Brazilian component of the Amazon rain forest has been lost in the past three decades. Overall, about two-thirds of rain forest destruction in the developing countries is caused by poverty-stricken small farmers who move to the forest margins and clear out trees to produce subsistence farms (many of which subsequently fail because of poor agricultural practices).[20] These poor farmers are not ecologically insensitive, but in their struggle for survival they view conservation of natural resources as less critical than their own short-term survival. The specific causes of deforestation continue to vary geographically. In Africa the cause has been mainly an increase in subsistence farming, with closed (undisturbed) forests yielding to shrubs and other land cover. In Asia the causes include changing cultivation practices by the rural population, government resettlement schemes, and large tree-plantation programs. In Latin America, especially Amazonian Brazil, the cause has been mainly conversion to other land cover by centrally planned operations such as government resettlement schemes and hydroelectric reservoirs, as well as government-subsidized cattle ranches.

There are signs that the situation may be improving. According to the United Nations, the rate of forest loss in developing countries "appears to have slowed somewhat since the last decade." The UN data show an annual rate of loss 12 percent smaller in the 1990–1995 period than in the 1980–1990 period.[21] And in the Amazon forest, according to Brazilian satellite data, the rate of deforestation in 1997–1998 was about 15 percent lower than the average rate from 1988 to 1996.[22]

It is important to note that the practice of deforestation is not a recent phenomenon but goes back centuries to the indigenous populations, who modified their natural environment in many ways. One cannot, of course, predict when the centuries-old practice of deforestation will actually be reversed, but one can point to conditions that are most likely to save the rain forests. First and most important is the fostering of efficient and environmentally sound agricultural practices that use natural resources to boost livelihood security without depleting those resources. It is imperative that small farmers be helped to intensify their agriculture, whereby high-value agriforestry and perennial crops are favored over fire-maintained cattle pastures and slash-and-burn farming plots.[23] Also critical are access to well-integrated, reliable markets, increased availability of credit, and roads that remain open all year.[24] It is also important that governments eliminate subsidies that have promoted farming of deforested land.

Economic growth may also be a key ingredient of saving the forests. Current improvements suggest that increased education and investment opportunities arising from economic growth are beginning to stimulate the application of efficient agricultural practices that include modern irrigation systems, new technologies, and better access to markets. In most cases, efficient agriculture will reduce or eliminate the need for new agricultural land and the incentive to clear marginal forest areas, and in favorable situations may even allow some existing agricultural land to be returned to forest growth (as is happening in the affluent countries). Further, to take pressure off land, nonagricultural sectors of rural economies also need to be strengthened through technological innovation, to assure higher productivity of both land and labor. A very positive sign is that timber firms in Brazil have discovered that sustainable management and reduced-impact logging of the forests can be at least 10 percent more profitable than the reckless conventional methods of timber extraction.[25] And not least, higher economic valuation of the rain forests can be achieved through their immense potential as ecotouristic destinations, living laboratories for medical and pharmaceutical research, and perhaps most important, cultural treasures that would take centuries to replace.

Despite these encouraging signs, one should be cautious when generalizing about the positive impacts of economic growth on forest conservation in developing countries. Unfortunately in a number of cases economic incentives are still working in the wrong direction. For example, some large-scale logging operations are still lucrative in the Congo and in Burma.[26] In Brazil some large soybean plantations have invaded the forest,[27] and in the Amazon region overzealous government development projects may be moving too rapidly in building infrastructure and opening land for colonization.[28] Economic development can be achieved in a manner consistent with preserving the world's great rain forests, but realizing this goal will require substantial investments by the international community in both the planning and financing of resource-conserving development projects. Worldwide the trend appears to be in the right direction, and this should inspire confidence that the road to affluence in the developing world is consistent with forest preservation, as has been the case throughout the affluent world. The price is high but worth it.

A critical issue here is time. Although the rain forests themselves will recover, as have forests everywhere, scientists presently have only rudimentary knowledge about the dynamics of the ecosystems associated with forests. Science cannot confidently predict the extent to which forest ecosystems may be altered by further forest reductions. Nor can science confidently tell us what the biological consequences of such alterations might be or the time required for the ecosystems' recovery, if they recover at all. These are areas where research is of the utmost importance. But research is not enough. A greater sense of urgency needs to be developed by the affluent nations as they consider policies and actions to help the developing nations preserve their rain forests.

COAL

Coal is the world's most abundant fossil fuel. From a strictly geological viewpoint, the supply of coal is so enormous (over ten thousand trillion tons)[29] that it would last for two thousand years even if the world continued to consume coal at today's rate, which is unlikely. The future use of coal, however, will not be determined by its geological abundance but rather by the market for coal in a world of increasing environmental constraints.

Once called "black gold," coal truly provided the energy resource base of the industrial revolution. In nineteenth-century Britain and the United States, coal was readily at hand, seemingly limitless in abundance and convenient to mine, transport, and use. Because of its many advantages over wood, coal had largely displaced wood as an energy source by the middle of the nineteenth century. Coal was one of the most important catalysts for the emergence of manufacturing societies throughout Europe and North America. Figure 19 shows the dynamic of U.S. coal use from the beginnings of the industrial revolution to the present (oil and wood use are also shown).[30]

But coal had its downside. It is not a clean fuel. Mining coal is inherently messy and hazardous, and burning coal is inherently dirty and polluting. As the new industrial cities were developing in Britain, terrible scars began to appear in the rural areas around coal mines, and entire regions became covered with dense smoke from coal-burning factories. But this was pollution with a purpose. Recall the industrialist in Charles Dickens's novel *Hard Times* who praised industrial smoke as "the healthiest thing in the world."[31]

During this period of industrialization, miners, factory workers, and their families were affected not only by visual ugliness from coal mining and burning but, more important, by a variety of health impacts, especially lung diseases caused by inhaling coal smoke. Nonetheless, the industrial societies tolerated coal-generated pollution for more than a century because it symbolized their jobs and their new prosperity. After World War II the huge industrial expansion brought on by years of unfilled consumer demand created a surge of affluence in the industrial countries, especially the United States. This affluence generated a renewed interest in the

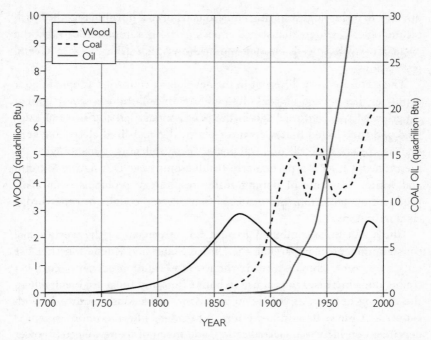

Figure 19. Dynamics of U.S. wood, coal, and oil use (1700 to the present), showing how the coal use displaced wood use after about 1900. Data are from Energy Information Administration, *Annual Energy Review* (Washington, DC: U.S. Department of Energy, 2000), tables F1a and F1b.

environment. People were living better and they wanted their surroundings to reflect their improving quality of life. Attention was focused on air pollution, which was as ubiquitous as ever but now becoming much more annoying.

A major target of the new attention was coal-burning power plants, whose sooty emissions blanketed wide areas of the country's broad industrial midsection. Eliminating coal-generated air pollution became a prime focus of national clean air legislation, including the Clean Air Act of 1970 and its amendments in 1990. Since 1970 dramatic reductions in sulfur oxides and particulate emissions from coal combustion have been achieved through government-mandated installation of "scrubbing" devices in the effluent stacks (once called "smokestacks") of coal-burning power plants. (See Chapter 7 for quantitative data.) The success achieved in recent years in reducing urban air pollution in the United States derives from a broad combination of public policies and private-sector investments, mostly directed to technologies cutting emissions from coal burning together with technologies cutting exhaust-gas emissions from automobiles. Today the air in most U.S. cities is cleaner and healthier than it has been in several generations.

The picture is very different in the developing countries. People in poor countries that depend on coal often endure air pollution as severe as that experienced two centuries ago in the newly industrializing cities of England and the United States. In most places the coal-fired electricity generators are old and still use antiquated "dirty-burning" technologies. As mentioned, China is a particularly troublesome case. During the autumn and winter, residents of Beijing suffer respiratory problems from coal-produced air pollution, and visitors often develop coughs or other bronchial irritations.

China and India combined account for 34 percent of the world's coal consumption. For comparison, the United States and Russia together use only 25 percent. China is already the world's leading producer of coal, yet tremendous increases in coal use in both China and India are inevitable in the coming decades, as coal is an abundant indigenous resource in both countries. China's determined effort to become a global economic giant is dependent on increased electrification, and much of its new electric power capacity will come from coal-fired generation. This huge growth in coal burning poses a significant environmental challenge to the global community, as neither China nor India is likely in the near future to enact or enforce environmental standards strict enough to bring about wide application of costly, state-of-the-art clean-burning coal technologies. In both countries the issue is more political than technical, because their expanded power capacity could be made both technically efficient and environmentally acceptable if investments were made in advanced generating technologies that are now commercially viable in the West. But China and India still see investment of scarce internal funds to meet Western air-quality standards as a rich country's luxury that they can ill afford. "You try to tell the people of Beijing that they can't buy a car or an air-conditioner because of the global climate-change issue. It is just as hot in Beijing as it is in Washington, D.C.," said Li Junfeng of China's State Planning Commission in an interview with the *New York Times*.[32]

Both China and India face an endless list of investment demands aimed at generating the economic growth that their people expect, including infrastructure such as highways and railroads and modern industries to compete for export markets with the developed world. These kinds of demands are much higher on national priority lists in most developing countries than are investments in environmental protection. Officials in China and other developing countries openly suggest that the developed countries should help pay for cleaner coal-burning technologies and other energy technology advances in the developing countries since the developed countries have been the main beneficiaries of past industrial development and are still responsible for most of the world's pollution.

Should we be optimistic about the ability of China and India to attract the huge foreign investments that will be required to ensure clean coal burning in the coming decades? Definitely so in the case of China, which is increasingly open to foreign investment, but less so for India, whose cumbersome social and governmental infrastructure needs a thorough overhaul.[33] In both cases development of environmentally clean electricity from coal is a critical challenge in the transition from poverty to affluence. With the huge profits that are possible, development of clean electricity from coal should be a win–win situation in both countries. And considering the huge size, population, and projected energy use of these two countries, ensuring the clean use of coal in China and India, as well as in other developing countries, would be a win–win situation for the entire world.

OIL

The media constantly remind us that oil is a depletable resource, that its supply is finite. Ever since the 1973 energy crisis a recurrent theme has been that the world will soon be running out of oil. Indeed, oil geologists periodically predict severe shortages of oil[34] and are generally pessimistic about future supplies.[35] The historical fact is, however, that even as the global use of oil continuously rose over the last century, the available supply kept *increasing* rather than *decreasing*. And in recent years the price of oil dropped to historic lows—a sure sign that supply is abundant relative to demand. How is it that the available supply of a resource can increase while it is being consumed?

What determines available supply is not the total amount of oil in the ground but the amount that can actually be extracted at a cost that allows marketing the oil under prevailing market conditions. Because the technologies of oil exploration and drilling constantly improve through research and development (R&D), production costs continually decrease. These improvements allow oil to be extracted economically from deposits that are ever more remote and more complex geologically. Thus the amount of economically available oil (the "reserves") can increase while the total amount of oil in the ground (the "resource") is actually decreasing. World

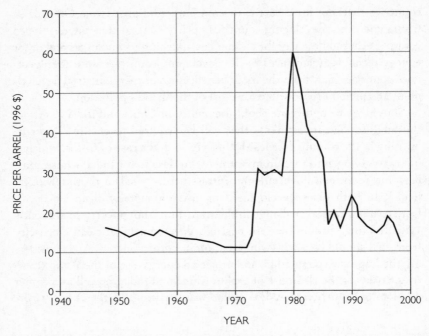

Figure 20.   Annual average crude-oil prices (1948–1998). Prices are inflation adjusted to 1996 dollars. Data are from *Energy Economics Newsletter* (WTRG Economics, n.d.), Web site at www.wtrg.com/oil.

oil reserves have actually been increasing faster than world oil consumption and are now at an all-time high.[36] Figure 21 shows the data from 1981 to 1993 for cumulative world oil production and identified oil reserves, demonstrating that during this period 254 billion barrels were consumed while 379 billion additional barrels were identified. If this trend continues, as is likely, there will be no shortage of oil in the foreseeable future.

But can the trend to increasing oil reserves continue? Billions of automobiles will be on the roads in the twenty-first century, especially if affluence increases as expected in the developing countries. Will the world's burgeoning automobile population not cause the demand for oil to outstrip the expected increases in production efficiency and drive up the cost of oil as the less expensive supply sources dry up? That is possible but unlikely, for two reasons.

First, great increases in the efficiency of conventional automobiles are likely (although these increases are not happening as fast as they should). The average fuel efficiency of the U.S. automobile fleet has increased from

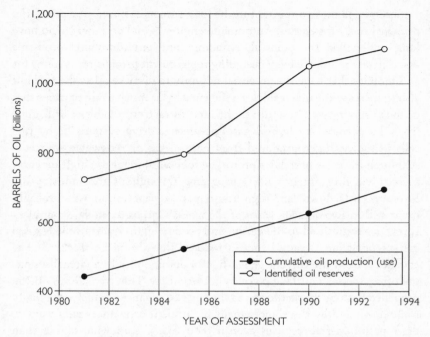

Figure 21. Cumulative world oil use and trends of identified (discovered) oil reserves. Data are from C. D. Masters and others, quoted in U.S. Geological Survey, *Changing Perceptions of World Oil and Gas Resources as Shown by Recent USGS Petroleum Assessments*, fact sheet FS-145-97 (Washington, DC, 1997).

18 miles per gallon (mpg) to 28 mpg since the auto fuel-efficiency standards were passed by Congress in 1975. In that legislation, light trucks were required to reach an average fuel efficiency of only 20.7 mpg. Because sport utility vehicles (SUVs) are classified as light trucks, the huge increase in their popularity has resulted in a loss of overall efficiency of the U.S. light-duty fleet (autos and light trucks), from 26 mpg (1987) to 24 mpg (2000). There is no technical reason why the fuel efficiency of SUVs and light trucks could not be much higher. It is inexcusable, in my judgment, that Congress has not yet extended the auto efficiency standards to include SUVs and light trucks. Energy efficiency expert Amory Lovins and others stress that technological advances in fuel and combustion chemistry, in transmission and drivetrain design, and in lightweight materials will double the fuel efficiency (miles per gallon) of gasoline-powered automobiles over the next decade or two.[37] As the number of cars increases, the total oil demand may rise only slowly and could actually decrease.

Second and more important in the long run, oil is on its way out as the principal fuel for vehicles. Automobile engineers and designers, who have long understood the political, economic, and environmental problems associated with oil, are busy designing replacement propulsion systems for automobiles that have the potential not only to displace oil as the fuel but also to increase the new vehicles' efficiency by as much as three times that of today's vehicles. These ultralight (and ultrastrong) vehicles will most likely be powered by hybrid gasoline-electric drive systems or by fuel cells using hydrogen produced from natural gas or renewable sources.[38] Although the costs of today's prototype fuel cells are far too high for commercial viability, intense R&D programs throughout the industry are bringing costs down, and significant market penetration will probably occur within two decades. Most of the world's large automobile manufacturers are committed to the design and production of hybrid vehicles, and early models are already being introduced into the niche market.[39] The transition from the current generation of automobiles to a virtually nonpolluting species will take place by mid-century. This transition is being propelled both by market forces as the costs of advanced vehicles gradually come down and by the choices made by affluent consumers who want to drive pollution-free vehicles even if their cost is somewhat higher than conventional vehicles.

In the next two decades, oil demand will surely continue to grow as the number of conventional automobiles burgeons in the developing world. But toward mid-century, unlikely as it may seem today, the fate of much of the world's oil may be to remain in the ground because of reduced consumer demand. When countries holding the world's major oil deposits, including Saudi Arabia and Iraq, begin to perceive a steadily diminishing future market for their oil, they may have their own "oil crisis" and thereafter engage in aggressive oil production and price-cutting to unload the oil while they can. Meanwhile, as long as demand remains high, the oil-producing countries will do their utmost to exercise their collective market power to keep a lid on production, so that prices remain high enough to satisfy their profit desires while still remaining acceptable to most consumers. The effects of recent OPEC production caps have been clearly felt by the U.S. oil-consuming public, which has seen the retail prices of heating oil and gasoline rise by about 50 percent in the last several years. This higher U.S. price level is likely to remain at least through the current decade, in spite of efforts to increase domestic oil production in Alaska and offshore.

NATURAL GAS

Among the fossil fuels, natural gas is the fuel of choice. It is abundant, it burns more cleanly than coal or oil with virtually no emissions of sulfur dioxide or particulate matter, and—of interest to those who obsess over carbon dioxide—it emits far less of the stuff (per unit of heat) than coal or oil.

As with oil, natural gas resources are very unevenly distributed around the world. The United States has only about 3 percent of the world's identified natural-gas reserves, whereas Russia is blessed with about 33 percent. Middle Eastern countries have 36 percent, a good part of which is in Iran. Israel has neither oil nor natural gas.

As with oil, the amount of the world's natural gas reserves keeps growing even as the use of gas increases, and for the same reason: continuing technological advances in exploration and production. An example of recent innovation is the successful production of gas from deepwater environments in the Gulf of Mexico, a technique not even contemplated twenty years ago. In 1975 the known world reserves of conventional gas sources were estimated at 2,348 tcf (trillion cubic feet),[40] whereas in 1999 the estimate had grown to 5,145 tcf.[41] The world's annual consumption of natural gas in 1998 was about 82 tcf.[42] Although it is tempting to divide 5,145 by 82 and conclude that the world's conventional gas supply will last sixty-three years, this arithmetic is incorrect because both the future reserve estimates and consumption rates will change (upward, no doubt) in unpredictable ways. One recent analysis concluded that conventional gas sources could last for almost two hundred years at current production rates.[43] There are also "unconventional" sources of natural gas (analogous to low-grade ores) that are probably larger than the conventional sources. These have not yet been tapped because of high production costs, but the costs will surely drop as new exploration and recovery technologies are developed during this century. All told, it is reasonable to expect that natural gas will be a viable energy resource for several centuries.

A number of developing countries are well endowed with natural gas deposits. For example, countries on the African continent (mainly Algeria, Egypt, Libya, and Nigeria) have identified reserves twice those of the United States. Malaysia, Indonesia, and China also have substantial reserves. Because of natural gas's superior qualities as a fuel, utilizing these reserves as these countries develop would contribute to both economic efficiency and outdoor air quality. However, exploring, extracting, and distributing natural gas requires a very costly and capital-intensive infrastructure, including production wells, processing plants, long-distance and local pipelines, and compressors to move the gas through the system. The United States, for example, has a 1.3 million–mile gas transmission and distribution system valued at nearly $150 billion.[44] Massive investments will be required to allow the developing countries to take advantage of natural gas's benefits as an energy source. Assisting these countries to develop clean energy sources is among the most important investments in the global environmental future that the affluent societies can make.

For electricity generation and many industrial fuel uses, natural gas is environmentally superior to other fossil fuels because it burns cleaner. When natural gas is burned, it produces virtually no atmospheric emissions of sulfur oxides and particulate matter and far lower levels of carbon monoxide and nitrogen oxides than combustion of coal or oil. It also produces almost no solid wastes. Because natural gas is a "low carbon" fuel, it produces smaller amounts of greenhouse gases (carbon dioxide) than coal or oil. In addition, most types of appliances and equipment that operate on natural gas are highly fuel efficient, and this contributes to lower environmental impacts. On the negative side, use of natural gas for heating and cooking may contribute to elevated levels of indoor air pollution (partly from unburned fuel) in comparison with electric appliances, which produce virtually no pollution (at the point of use). This can be a significant issue for people with asthma, allergies, or other respiratory problems.[45]

THE BOTTOM LINE ON FOSSIL FUELS

The world's fossil fuel supplies are plentiful. They will neither run out nor become scarce in the foreseeable future. In the early days of industrialization, extracting and burning fossil fuels—first wood, then coal, then oil—did make a sorry mess of the environment even as they brought affluence to the fossil-fuel users. As citizens of the industrializing societies achieved greater freedom and affluence, they decided to restore their environment, and have been remarkably successful in doing so, by sanctioning increasingly strict environmental policies and supporting development of technologies for cleaner burning of fossil fuels. In contrast, the developing countries are still typically in the early industrializing, and polluting, phase of fossil fuel use. But recently the global-warming issue has thrown a huge roadblock in the path of continuing growth in use of fossil fuels by both developing and developed countries. Although the causes and importance of global warming will probably remain a matter of scientific controversy for many years, the ability of fossil fuels to contribute to the world's economic development in the future as they have in the past has been thrown into some doubt. In the end, will fossil fuels be seen as the genie of global economic and social development or as the culprit of global environmental degradation? And what are the alternatives to fossil fuels? For the most part, renewable energy sources and nuclear power. Can these replace or sufficiently augment fossil fuels? Read on.

9

SOLAR POWER TO THE PEOPLE

If you are over forty, you certainly remember the long and frustrating lines at gasoline stations during the so-called energy crisis of 1973–1974. So dramatic were the fuel shortages and pessimistic media reports that many consumers believed the world was actually running out of oil. For the first time people became concerned that their energy-dependent life-styles might be in jeopardy. In hindsight we now know that the gasoline queues had more to do with the monopoly market power of OPEC oil pro-ducers than with resource scarcity. There was in fact no scarcity of energy resources. Yet, by deliberately reducing their oil production and causing artificial (though short-lived) shortages, the oil producers were able to create chaos in the consumer countries, drive up the price of oil products everywhere, and most important, forcefully challenge the widespread (some would say, sacred) notion held by consumers in affluent countries that energy would always be available and cheap.

Although environmentalists decried OPEC's political power, generally they welcomed the prospect of higher fuel prices. Already concerned about the environmental impacts of fossil fuel use, environmentalists saw the oil price increases not as a lifestyle threat but as an opportunity—an over-due economic incentive for society to reduce its use of fossil fuels. The term *energy conservation* entered the popular environmental vocabulary of the 1970s. Exhorting his fellow citizens to save energy, President Jimmy Carter referred to energy conservation as "the moral equivalent of war" and appeared on television wearing a heavy woolen cardigan, presum-ably to demonstrate that the White House was being underheated to save energy.[1]

The higher gasoline and fuel-oil prices of the 1970s not only stimulated people to save energy but also encouraged interest in alternatives to fossil fuels, including resources that previously were too expensive to compete with fossil fuels. The nonfossil energy alternative that became most popular among environmentalists was the development of "renewable resources." Mostly derived from the sun's energy, renewable resources appeared to have a number of advantages over fossil fuels: they would be kind to the environment, they would use "free" energy from the sun, and they would be available forever (or at least until the sun burns out billions of years hence). That seemed to be an unbeatable combination. History too was on the side of renewable energy, since humankind's energy needs had been served for millennia by renewable resources such as wood, charcoal, and animal and plant wastes, long before coal and oil were discovered. In the 1970s these traditional renewable resources were still widely used in the poorest societies, and some are still in use today.

In preindustrial times the applications of renewable resources for cooking, heating, and metalworking were primitive and very inefficient. But in the 1970s, environmental technologists offered new variations of the old theme. By applying high-tech science and engineering to the traditional technologies, they embarked on the development of a new generation of renewable technologies that would be sophisticated enough to replace most of the current energy applications of fossil or nuclear fuels yet also environmentally benign and, in principle, sustainable indefinitely. During the decade following the 1973–1974 energy crisis, this "joint venture" between environmentalists and technologists in the affluent countries led to the introduction of many novel renewable technologies as well as to a resurgence of interest in hydroelectricity, a well-established renewable technology. The new renewables included upscale solar heating and cooling systems for residences and a variety of technologies for electricity generation ranging from high-tech windmills and solar heat "farms" to incinerators that turned urban wastes into electricity. Even the lowly wood-burning stove made a surprising comeback in the 1970s in contemporary, high-tech form.

Renewable energy soon became part of the environmentalist's credo. The term *soft* was introduced to connote the desirable characteristics attributed to solar and other renewable technologies in contrast to the undesirable characteristics attributed to the *hard* fossil or nuclear technologies.[2] The major attributes of soft energy technologies claimed by their advocates were perpetual availability, environmental superiority, and ideological compatibility with the new lifestyles. Another word joining the soft energy lexicon had a decidedly political origin. *Green* was originally coined by European political groups who vigorously opposed nuclear power while promoting strong government environmental controls and renewable energy technologies.

SOLAR ENERGY

The generic terms *solar energy* and *renewable energy* cover a wide variety of processes for converting the sun's energy into forms useful for heating, cooling, generating electricity, and producing fuels. Typically these processes collect the sun's energy either directly when the sun is shining, with use of large collectors (as in rooftop solar collectors for heating buildings), or indirectly (as in use of windmills, falling water, or biomass burning for electricity generation). In the 1970s these technologies, with the exception of hydro, were at an early stage of development and far from being commercially viable. But political support grew rapidly in the United States for development of alternatives to imported oil, and this led federal and state governments to initiate large R&D programs directed to developing renewable energy technologies and promoting their commercial application. Government support was not limited to R&D but soon included substantial subsidies and tax breaks selectively awarded to new commercial applications of renewable energy technologies, with the goal of assisting them to become economically competitive with traditional fossil fuels. Government programs were also established to transfer new renewable energy technologies to developing countries, where they were intended to supplant the traditional use of wood and wastes.

FREE BUT NOT CHEAP

It is important to keep in mind a fundamental difference between energy from the sun and energy from fossil or nuclear fuels. Solar energy falling on the earth's surface is extremely dilute (i.e., weak), in contrast to the much more concentrated energy content of fossil and nuclear fuels buried in the earth. Although the sun's heat is "free," it is spread so thinly over the earth's surface that a land area the size of Hawaii would be required to collect enough solar heat to provide the amount of electricity generated by only the currently operating U.S. nuclear power plants. The diluteness of the sun's energy reaching the earth is the main factor determining the cost of solar and other renewables. Developing the technologies required for concentrating the sun's rays to the point where useful energy can be extracted is a technically complex and expensive engineering problem.

This is a poorly understood but critical point that underlies the cost problems associated with solar energy.

Given this fundamental disadvantage, it was realized from the outset that the greatest obstacle faced by renewable energy technologies would be high cost. Fossil fuel technologies, including electricity generation with coal, oil, and natural gas, and the automobile, with its oil-fueled internal combustion engine, had not only the advantages inherent in their convenient and highly concentrated fuels but also the advantage of a century's technical development and commercial application. As a result, the fossil fuel technologies enjoyed a tremendous technical superiority and widespread market acceptance. Despite government R&D support and generous market subsidies, the renewable technologies had to play serious catch-up with an uncertain end point.

Nonetheless, in the crisis atmosphere of the 1970s the prospects for renewable energy seemed bright. A 1976 government study estimated that wind power could supply almost one-fifth of all U.S. electricity by 1995. In 1978 the Carter administration announced its goal for renewable energy sources to supply at least 20 percent of U.S. energy use by 2000.[3] The organizer of International Sun Day in 1978 went even further, saying "Forty percent of our energy could come from solar energy by the year 2000 if we make some dramatic moves now."[4]

Two decades later, it is difficult to be so sanguine about the future of renewable energy. On the positive side, the costs of some renewable technologies that were prohibitively high in the 1970s have come down significantly, and some applications have even surpassed the cost reduction targets set in the 1970s. For example, the cost of wind-generated electricity dropped from 55 cents per kilowatt-hour (kWh) in 1975 to 4–6 cents per kWh in 1995.[5] This cost, however, reflects about 2 cents per kWh of federal and state subsidies. At the subsidized price, wind power will become increasingly competitive with natural-gas generation if the price of natural gas continues to rise. (In 2000 natural-gas generation was averaging about 3.5 cents per kWh).

Still, in spite of the cost reductions and government subsidies, renewable energy technologies have proved to be much less successful in the commercial market than their advocates had projected. In total, renewable energy sources now supply only about 3.8 percent of U.S. energy consumption (about 6.9 percent if one includes hydroelectric power),[6] nowhere near the 20–40 percent targets referred to above. And by 1996 wind generation had captured only a 0.1 percent share of the U.S. electricity market, in contrast to the 20–30 percent share that had been projected in the 1970s.[7]

Why have the soft energy technologies thus far failed to compete in the market and become a viable complement to or replacement for fossil and nuclear energy? Can the rising tide of new technology lift the soft energy technologies as well, allowing them to achieve significant roles in the affluent nations? Will continuing government subsidies be required? What role will the soft technologies have in the developing nations? Let's look at the renewable energy technologies once considered most promising—hydroelectric generation, wind electricity generation, solar electricity generation, solar heating, geothermal energy, and biomass conversion—to see where the soft pathway is actually heading.

HYDROELECTRIC POWER

Hydroelectric power is the leading renewable energy source in the United States, producing 55 percent of the country's total renewable energy consumption and 98 percent of electricity obtained from renewables (1996 figures).[8] The usable energy of hydropower comes from falling water: dams store the water at a higher elevation and allow it to fall in a controlled way to a lower elevation, the water passing en route through turbines that generate electricity. Hydropower has long been a major source of electricity in countries that have mountainous regions with heavy rainfall located near population centers—for example, Canada, Norway, Sweden, and Switzerland. Worldwide there are more than forty thousand large dams, of which nineteen thousand are in China and fifty-five hundred are in the United States. These dams provide about 20 percent of the world's generated electricity.[9] In the United States, hydro provides only about 10 percent of the total generated electricity, though in California hydro's contribution is 22 percent (1998 figure).

Hydro is a truly renewable energy source because it depends on the earth's hydrological cycle, which recurs each year. The water stored in dams is continually replenished by rainfall (except in drought years). Hydro also conforms to the major environmental criteria for "soft" energy since it creates no air pollution and releases no carbon dioxide to the atmosphere. From the perspective of the climate change issue, hydro would appear to be an almost ideal power source since it emits no greenhouse gases (although the claim has been made that rotting vegetation in dam reservoirs can produce significant amounts of greenhouse gases[10]). In addition, the economics of hydro generation have been very favorable, with the generation costs for some existing hydro installations averaging less than half the costs of fossil fuel generation.

Nonetheless, in recent years hydro has fallen out of favor with mainstream environmentalists because dams have been found to have destructive impacts on fisheries, wetlands, forests, and aquatic life. Particular concern has been expressed about the damaging effects of hydroelectric facilities on aquatic life passing upstream, downstream, and through the sites.[11] Other concerns include loss of recreational areas to accommodate hydroelectric facilities, the potentially catastrophic effects of dam failures, and various health and ecological considerations.[12] Some of these impacts are being mitigated by development of new hydro technologies that reduce dams' impact on water quality and aquatic habitat and enhance the survival of fish passing through hydro turbines.[13]

Probably hydropower's most serious drawback has been the risk of catastrophic dam failures, a constant threat to the lives of humans living downstream. Dam collapses have caused over thirteen thousand deaths worldwide, not counting the 1975 dam collapse in China, which may have killed over two hundred thousand people. In developing countries this risk is compounded by other problems, including forced resettlement of people from inundated lands (over three million people have been displaced), potential for outbreaks of water-borne diseases, and intensification of regional water-rights conflicts.[14]

Because of these impacts, hydropower development has become so unpopular among environmentalists that they generally no longer refer to hydro as a renewable resource. This trend is illustrated by the promotional literature of a supplier of "green" electricity in California, which excludes "large hydroelectric" from its list of "100 percent Renewable Power" options and instead lumps hydro together with the nongreen options coal, natural gas, and nuclear power.[15] Another example of hydro's loss of favor is the criticism directed at the Clinton administration by the Sierra Club and Trout Unlimited when the administration promoted hydro in 1993 as part of a proposed strategy to curb global warming.[16] In view of the growing opposition from environmentalists, hydropower will probably decrease significantly in the future U.S. energy mix since no new hydro sites are likely to be developed and an increasing number of current sites will probably be shut down. The former head of the U.S. Bureau of Reclamation (which built most of the large U.S. dams) has declared, "The dam building era in the United States is now over."[17]

Opposition from environmentalists in the affluent countries notwithstanding, considerable expansion of hydropower is taking place in the developing world. Many prime hydro dam sites remain, and governments will probably continue to rank hydro's potential contributions to their nations' economic and social development far ahead of the kinds of environmental concerns that are of importance in the United States. A prime example is the monumental Three Gorges Dam project on the Yangzi River in China. When completed, this will be the largest dam in the world, creating the world's most populous inland port at Chongqing and providing capacity for electricity equivalent to about thirty coal-fired plants (18.2 million kilowatts, or about 10 percent of China's present electric power capacity). Almost 1.5 million people are being resettled to make room for the dam and its four-hundred-mile-long reservoir. Vigorous international objections have been expressed over the potential environmental impacts of this dam, including massive animal habitat destruction, river sedimentation, and the de facto population resettlement. Nonetheless the project is proceeding, and the Chinese government holds to its position that the clean electric power and long-overdue flood control provided by Three Gorges will more than compensate for any environmental liabilities. The dam is due for completion by the end of this decade.

Brazil provides a more typical perspective on the continuing hydropower expansion in the developing world. Brazil is almost totally dependent on hydro, which provides 87 percent of the country's total electric-power capacity. While Brazil's planned dams are far smaller than China's gigantic Three Gorges dam, their social impacts are no less significant. In Amazonia sixty-eight new dams are scheduled for completion by the year 2010, with a total of eighty new dams proposed for 2040. These dams will flood 2 percent of the region and displace about 250,000 people, around 60 percent of the local populations. Government-sponsored electric power projects such as these, which can adversely affect the cultures and livelihoods of so many people, have produced new social movements in Brazil aimed at securing the participation of affected citizens in the decision-making process. Generally these movements support government economic development programs and do not oppose hydroelectric development as such, but they demand that the government incorporate the goals of local and regional socioenvironmental sustainability in the planning process.[18]

While the evidence indicates that the use of hydroelectric power has reached its zenith and may even decline in the affluent countries, most of which are blessed with multiple energy options, hydro will probably remain the renewable energy technology of choice in the developing world for the foreseeable future. Although hydro facilities are costly to construct, once in place they can provide inexpensive and dependable electricity for decades.

WIND ENERGY

Even a casual visit to the "wind farm" at Altamont Pass in northern California leaves a strong visual impression of this vast electricity-generating installation, with its hundreds of whirring propeller-driven wind machines towering above the rolling hills. Descendants of traditional windmills, these wind power plants are actually high-tech turbines that convert the energy in the motion of wind to electrical energy.

With hydropower's fall from favor in the affluent countries, wind has now become the preferred resource for electricity generation among many environmentalists. According to a U.S. wind-energy industry organization, the worldwide installed electric capacity from wind has grown from under 2 million kilowatts (mkW) in 1990 to 13.4 mkW at the end of 1999. Germany leads with 30 percent of the world's installed capacity from wind, the United States is next at 19 percent, and Denmark and Spain follow at 13 percent.[19] In Denmark 10 percent of the country's total installed electric capacity comes from wind, but in the United States wind provides only about 0.1 percent of the total installed capacity.

As a renewable resource, wind power has much to commend it. The large wind farms can supply significant amounts of electricity to the main grid systems when the wind blows, while smaller turbines can be used by farms, homes, and businesses in windy locations, such as along coasts, and also can be used in remote areas to which bringing power lines would be prohibitively expensive. In off-grid locations that depend on wind-generated electricity, however, a local backup electricity source is required for periods when there is little or no wind.

Wind farms have some important environmental advantages: they require no fuel, consume no water, and emit no air pollutants, greenhouse gases, or toxic wastes. These attributes help to explain why wind energy is the most favored "green" energy source. But wind farms also have significant environmental liabilities. They consume great amounts of land, requiring about seventeen thousand acres to produce the electricity equivalent of one nuclear power plant. The turbines generate a rumbling propeller noise, an irritant to people living nearby. Because of the turbines' huge size (up to 160 feet tall with 130-foot blades), a group of wind machines can constitute a serious visual blight on an otherwise beautiful landscape.

A vexing problem for environmentalists is the hazard to birds posed by wind turbines in some locations, where kills have been reported among federally protected species including golden eagles and red-tailed hawks.[20]

To avoid jeopardizing long-term support for wind power by environmental groups, this risk needs to be quantified and potential sites surveyed carefully to assure that they are away from major migratory paths and from large raptor habitats. A recent proposal to build several hundred wind turbines off the coast of Nantucket Island (Cape Cod) has encountered strong opposition on grounds of both esthetics and risks to wildlife. A positive example is the agreement reached between the National Audubon Society and a commercial developer for relocating a planned wind-farm site in the Los Angeles area almost fifty miles away to avoid interfering with the flight patterns of the endangered California condor.[21]

Apart from these environmental issues, the most serious liability of wind power is cost. Although, as stated earlier, the cost has dropped dramatically since the 1970s, wind power still relies on government subsidies to be competitive with other electricity technologies. The present cost of wind generation, 4–6 cents per kilowatt hour (kWh), includes federal and state subsidies of approximately 2 cents per kWh.[22] For comparison, the operating cost (in 2000) of generation with natural gas is about 3.5 cents per kWh, coal about 2.1 cents per kWh, and nuclear power about 1.8 cents per kWh. Unless the cost of wind power decreases significantly or the price of natural gas rises significantly, wind will probably not become competitive in the electricity market in the absence of subsidies.[23]

Several factors contribute to wind power's relatively high cost. First, the intermittent and unpredictable nature of wind means that turbine equipment remains idle during periods of little wind, more than offsetting the fact that wind is free when it blows. Second, sites must be located in the windiest places, which often turn out to be remote from population centers, so that expensive transmission lines must be built. In contrast, natural-gas generation turbines can be located near points of use.[24]

Some advocates of wind generation acknowledge that only with continuing government subsidies can wind compete with fossil fuel technologies, but they argue that "wind energy has the potential to provide tremendous economic and environmental benefits."[25] Ultimately, consumers will decide by their actions in the marketplace whether those claimed benefits justify paying higher prices for wind energy at a time when supplies of natural gas, possibly the world's best fuel overall, appear to be sufficient for hundreds of years and when natural-gas technologies are continually improving.

One can argue that subsidies for wind power were critical in the past as a catalyst for developing a technology that the general market would not then have supported. As a result wind technology is now available for an important but small niche market—electricity generation at certain remote and windy sites where fuel delivery would be expensive and transmission-line extensions prohibitive. However, if government subsidies for wind power do not continue for long, as may be likely, the use of wind power will probably not expand much beyond that niche market.

SOLAR ENERGY

Although all renewable resources ultimately depend on the sun's energy, the term *solar energy* commonly refers to technologies that use the sun directly for generating electricity or heating and cooling buildings. Environmentalists favor solar technologies because they use no depletable fuels and their operation causes no air pollution. But these technologies also have environmental drawbacks—generating solar electricity to serve large markets, for example, requires very large land areas, imposes high construction and operating costs, and uses materials whose manufacture causes pollution.

Electricity can be generated from sunlight in various ways. One technology, called "solar thermal," uses a system of shaped collectors to concentrate the sun's rays, heating a fluid to operate a thermal engine that in turn drives electric generators. (This is more or less the same process as used in a fossil-fueled power plant, except that the source of heat is sunlight rather than fuel combustion.) Two large pilot plants of the solar-thermal type have been built in the United States, the more recent being a collaboration between the federal Department of Energy and a consortium of utilities. This plant operated for somewhat over a year, during which time it demonstrated that it could generate electricity steadily. The cost of the electricity it produced was never publicly announced, but it was clearly several times higher than that of fossil fuel generation. A land area of about four million acres—the size of Hawaii—would be required for central solar plants of this type to produce electricity equivalent to the country's one hundred or so nuclear plants. One study otherwise very sympathetic to renewable energy concluded that "the future of all the central solar generators is in doubt."[26]

A more promising solar technology for electricity generation is "photovoltaic," which uses panels of large solar cells similar to the small cells found in batteryless pocket calculators. Although costs are already well below those of solar-thermal electricity generation, they are still a long way from making solar-photovoltaic competitive with natural-gas electricity generation. Optimism may be in order for the long term, however, because R&D is vigorous in this field in both public and private sectors and technological breakthroughs in design and composition of photovoltaic cells are likely. The land area required for large-scale photovoltaic farms is probably less than half that of solar-thermal farms, though still very large. And there are other environmental liabilities. Huge photovoltaic farms, like thermal farms, create visual blight and some change in local climate. Large-scale use of photovoltaics also poses the risk of chemical pollution from the huge amounts of toxic materials, including arsenic, cadmium, and gallium, that are used in the manufacture of photovoltaic cells.

WHY IS RENEWABLE ENERGY SO EXPENSIVE?

People often ask why the costs of renewable energy technologies have turned out to be so much higher than fossil energy in spite of continuing heavy public and private R&D investment. Clearly these high costs have seriously inhibited the renewables' market penetration, at least for the near term, and have dampened the enthusiasm for renewable energy among all but the most dedicated advocates.

History provides an unambiguous answer. Recall that most of the R&D work on renewable technologies was begun during the crisis atmosphere of the 1970s. The overriding factor stimulating the government's crash R&D programs on renewables was the almost unanimous prediction among energy experts, especially after the second oil price hike in 1979, that the world price of oil would continue to rise beyond $30–$40 a barrel (the highest point in 1980 and in 2000) and would shortly exceed $100 a barrel. Most technologists believed, and reasonably so, that at the $100 price level the renewables could eventually compete with fossil fuels. Of course, what actually happened was quite the opposite: after 1980, oil prices did not continue to rise but instead dropped precipitously to below $10 a barrel. Even though oil prices rose again to the $25–$30 range in the late 1990s, this is still far below (in inflation-adjusted terms) the price levels predicted two decades earlier. Today, oil prices appear to be stable and will probably remain so for decades, while natural gas prices may actually fall in the future owing to new discoveries and technological advances. Although the costs of renewable technologies are also continuing to come down as the result of technological advance, the cost gap between renewables and fossil fuels remains large, and there is little prospect for renewable energy to become competitive with fossil energy in the near term. Yet, encouraged by government subsidies and R&D, private investors have poured billions of dollars of capital into large-scale renewable energy technologies and facilities. Today these enterprises find themselves in a difficult competitive situation, and it is probably only the ideological devotion of their advocates and the continuation of subsidies that keep them alive. How long this situation will continue is anyone's guess.

RENEWABLE ENERGY IN DEVELOPING COUNTRIES

In the developing countries the situation with renewables is more promising but also more complex. Nowhere is the dichotomy between the environmental problems of the poor and the rich nations more apparent than in the case of renewable energy. The affluent societies continue to invest in large-scale renewable technologies that they don't need, while over two billion people in the developing world lack even the basic energy services of electricity and heat, and their desperate need for small-scale and efficient energy technologies goes unfilled. Renewable energy resources could play a major role in filling this need.

Most rural families in the developing countries still rely on kerosene and fuel wood for their light and heat. The heat from their open wood stoves is accompanied by toxic smoke permeating the small, enclosed living spaces. Without electricity the nights are dark and parents cannot engage in useful activities, nor can children do schoolwork. In recent decades, many developing-country governments have undertaken large-scale conventional electrification programs, but typically they have found that, even with the help of international loans, they cannot afford the huge capital costs of building large power plants or running transmission lines to the thousands of outlying villages. For these reasons many large electrification schemes have failed.

Smaller-scale renewable energy technologies are a much better fit to the needs of rural households, especially those in remote, isolated, and ecologically fragile areas. The technologies that appear to be most promising in the near term are solar-photovoltaic utilities, particularly for remote islands and villages; minihydropower plants, where adequate sites are available; wind-turbine generators, where there are favorable wind sites; and enhanced use of biomass fuel where adequate sustainable biomass resources exist. Renewable sources are also particularly suitable for providing energy to populations living in environmentally fragile areas, such as small islands, deserts, river deltas, and high-mountain zones. A 1995 workshop on household solar-power systems concluded that these

> represent a clean, climate-friendly alternative for rural electrification. During the past five years, remarkable advances have been made in the economics and technology of solar cells. Costs have declined by more

than two-thirds, and solar-cell efficiency has more than doubled. Given these improvements, the widespread use of household solar units (which can operate several fluorescent lights, a television and a small appliance for up to four hours) is now a viable option. Solar photovoltaic units are cost-effective relative to other available energy sources, far cheaper than grid extension, and profitable for companies to provide. Model projects in several Asian countries and the Caribbean have shown that demand for these systems is high and that rural households can afford them if financing is available.[27]

In spite of these benefits, private entrepreneurs have generally not taken advantage of the opportunity represented by the millions of developing-world households that need and could buy these renewable systems. A major obstacle has been the lack of suitable market infrastructures capable of handling the required capital flows. The current infrastructures were developed around construction projects for multimillion-dollar power installations, which rely on single-point lending and investment for indi-vidual large projects. These infrastructures are rarely appropriate for financing the purchases of small, inexpensive solar systems by millions of widely dispersed rural households. Although model projects have demon-strated several appropriate delivery mechanisms for getting credit and solar household units to rural end users, these relatively modest success stories have not been sufficient to raise the confidence of traditional investors.[28]

Thus the situation with renewable energy displays the same unfortu-nate dichotomy between rich and poor that characterizes most of today's environmental situations. The rich countries, influenced by Green politics and aggressive environmental public relations, continue to provide public subsidies for investments in large-scale renewable technologies for which there is little need and which, in any case, are neither economically com-petitive nor environmentally superior. Although still ideologically popular, in the end these technologies will probably fail because they do not meet the true tests of the competitive market.

In contrast, the poor countries have a tremendous need for renewable energy sources, and a number of ingenious yet affordable technologies have been available for years. Not only the lack of infrastructure but also a woeful lack of domestic political support inhibits the development of a viable market for renewables in many countries, and the result is that sev-eral billion people remain without the basic energy services they need and to which they have a human right.

A comprehensive solution for providing renewable energy to the devel-oping world cannot easily be found because the histories, cultures, and political structures of developing countries are so varied. Yet there are common themes. Most pervasive is the lack of dedication on the part of government leaders in many developing countries, especially in Africa, to elevating their people out of poverty. Without such dedication, the growth of indigenous infrastructures for developing renewable energy and other appropriate technologies will remain only a distant goal, along with many other critical development goals.

10

NUKES TO THE RESCUE?

Environmentalists in the affluent countries are faced with a perplexing dilemma: how to choose between fossil fuels and nuclear power for future electricity production. Setting aside the option of renewables, which are unlikely to contribute more than a small fraction of the world's electricity in the foreseeable future, one might describe the dilemma as a "Sophie's choice" because, to most environmentalists, neither of the two realistic alternatives is acceptable. On one hand, fossil fuels—coal, oil, natural gas—produce greenhouse gases, which most environmentalists associate with global warming. On the other hand, nuclear power is rejected as technologically unsafe and socially inappropriate.

This dilemma notwithstanding, nuclear power continues to be a major player in the world's electricity system. In 1999 commercial nuclear power plants supplied 75 percent of the electricity in France, 47 percent in Sweden, 36 percent in Japan, 31 percent in Germany, 27 percent in Britain, and 20 percent in the United States.[1] From the perspective of future *availability* of energy resources, nuclear power is a strong option, stronger than fossil fuels. Even with current reactor systems, the amount of terrestrial uranium that is "reasonably assured" could fuel nuclear power plants for many hundreds, if not thousands, of years.[2] And the uranium contained in seawater alone could supply the entire world's electricity for many thousands of years.[3] If, in addition, advanced "breeder" reactors become technically and economically practical, the nuclear fuel resource would be so abundant that it could be considered essentially renewable.[4]

In spite of its immense potential as an electricity source, nuclear power has been swathed in controversy ever since its introduction in the 1950s.

Many environmental groups and Green political parties had their beginnings as participants in the antinuclear movements of the 1960s and 1970s. During that period the American public developed a strong perception of nuclear power as an unsafe technology,[5] and that sentiment largely persists despite U.S. nuclear plants' long record of safe operation.[6] As a result of public opposition, some nuclear plants that have reliably produced electricity for decades are being dismantled years ahead of the end of their useful lives. Antinuclear sentiment is even stronger in Sweden and Germany, and both countries are committed to the complete phaseout of nuclear-generated electricity over the next several decades. In contrast, France and Japan will almost certainly continue to depend on nuclear energy for the foreseeable future.

Technical issues alone do not fully explain the American public's anxiety about nuclear power. Over four decades, commercial nuclear power has had an excellent safety record worldwide, with the notable exception of a system used in the former Soviet Union (FSU) and some former Soviet-bloc countries. History's worst reactor accident, which occurred at Chernobyl in the FSU in 1986, claimed thirty-one deaths from radiation exposure in the short term and is expected to cause an approximately 0.3 percent increase in the normal cancer death rate among the exposed population.[7] That accident was the product of an atrocious Soviet reactor and plant design (RBMK-1000) that would never have been approved for commercial use in any Western country.

Outside the former Soviet Union, about eighty-five hundred reactor-years of operation have taken place with no accident involving a large external release of radioactivity.[8] The United States' only serious reactor accident, at Three Mile Island, Pennsylvania, in 1979, released insignificant amounts of radioactivity and caused neither deaths nor injuries from radiation. The main short-term health impact of that accident was severe psychological trauma to local residents fanned by alarmist media coverage and inept government crisis management. Over time, there have been many occasions when reactor systems malfunctioned in one way or another, but their built-in safety systems have worked reliably enough to prevent serious consequences. The popular perception that any incident or malfunction at a nuclear power plant will produce a major disaster is simply incorrect.

NUCLEAR FEAR

Historian Spencer Weart has suggested that the public's negative attitude toward nuclear power is grounded mostly in fear. He argues in his book

Nuclear Fear that the antipathy stems partly from an age-old association of radiation with wizardry and the supernatural. But more important, the fear arises from a misplaced association between the "war atom" and the "peace atom."[9] The war atom relates, first, to the United States' first use of nuclear bombs against Japan in 1945, and, second, to the ever present threat during the Cold War of a hydrogen bomb exchange between the United States and the Soviet Union—a threat of holocaust proportions that ended only with the end of the Cold War in 1991. In contrast, the peace atom refers to the commercial generation of electricity—a civilian application of nuclear technology only remotely related to military weapons applications. The first commercial nuclear electricity plant began operation in Shippingport, Pennsylvania, in 1957, more than a decade after the end of World War II. The Pennsylvania location was fortuitous yet appropriate, since it was there during the decade of the 1950s that the United States' worst smog episodes from coal burning were experienced.

THE NUCLEAR DEBATE

In the United States the thirty-year nuclear debate has failed to produce a clear distinction in the public mind between the industrial-level safety issues associated with the commercial use of nuclear power and the grave national-security issues posed by proliferation of nuclear weapons around the world. Critics of nuclear power have never accepted such a distinction and have argued that the civilian and military issues are inseparable because development of civilian nuclear-power technology, even if completely safe in operation, could facilitate a country's development of nuclear weapons. The opposing view is that any country deciding to develop nuclear weapons will most likely do so through a dedicated program rather than piggyback-ing on commercial nuclear power. To date, commercial nuclear power has played little, if any, role as a catalyst for any country's entry into the nuclear arms race, nor are there any known cases in which individuals or subnational groups have stolen materials from nuclear power facilities for use in weapons.[10] Nonetheless, the two issues have remained tightly linked in the public mind—so tightly, in fact, that the expression *Nuke* is commonly used to refer to both civilian electricity plants and hydrogen bombs.

With the end of the Cold War the specter of military use of nuclear weapons by the major powers has practically vanished. Russian and American officers were actually stationed at each other's primary missile sites to be certain that no accidental launches could occur during the most critical moments of the "Y2K" computer changeover at the start of the new millennium. Yet even as the world breathes easier at the demise of the Cold War nuclear terror, new threats of nuclear violence, exemplified by the nuclear weapons competition between India and Pakistan, are on the rise, and terrorist attacks employing crude nuclear devices or makeshift radioactive ("dirty") bombs are increasingly likely. *Nuke* is still very much a pejorative term in the American vocabulary.

Yet military dangers and associations are not the only reason why antipathy toward commercial nuclear power remains. Another important factor is distrust—a widespread belief among Americans that their government has been less than forthcoming in providing accurate information about the safety (or risks) of nuclear power. During the Cold War, national security obviously required secrecy about U.S. nuclear weapons testing, but the distinction between the significant risks to the public associated with testing—for example, the worldwide radioactive fallout from tests—and the much smaller risks of commercial nuclear power were never explained convincingly to the public.[11] With the clarity of hindsight it is apparent that the government could have done a much better job in clarifying this distinction. At any rate, the legacy of distrust lingers on, not only in matters relating to nuclear power but more broadly as well.

RADIATION

Another element of the public's nuclear fear centers on exposure to radiation, even at extremely low levels. Although little risk is associated with low-level radiation, unfortunately neither the government nor the scientific community has established realistic guidelines about low-level radiation exposure and communicated these clearly to the public. It is well-known that *high-level* radiation exposure is dangerous and often fatal (the death of many atomic-bomb victims from radiation sickness demonstrated that). But no health hazard has been observed from exposure to radiation at very low levels—for example, routine medical and dental X rays, the earth's natural radiation background, or the very low levels near nuclear power plants. For an analogy, think about sleeping pills—if one person ingests fifty sleeping pills (high-level exposure) the result would almost certainly be fatal, but if fifty people ingest one sleeping pill each (low-level exposure), would one person die? Very unlikely. Yet this type of *linear* assumption underlies the current government radiation standards, which treat radiation exposure as being proportionally dangerous all the way from high exposure down to zero exposure. But there is no empirical evidence that radiation exposure at very low levels carries a significant risk. Some experts even believe that low-level radiation exposure may be beneficial to health, although there is no empirical evidence to support this belief either. Yet in view of the demonization of all nuclear radiation, regardless of level, is it any wonder that many people fear being in the proximity of a normally operating nuclear-power plant with its minuscule radiation level?

A COMEBACK FOR NUCLEAR POWER?

Recent polls notwithstanding, so deeply is nuclear fear entrenched in the public mind that another generation is likely to pass before a significant comeback of nuclear power could be a realistic scenario for the United States. Should the current concern about global warming continue, however, nuclear electricity's resurgence may be hastened by the advantage it enjoys over fossil fuel plants in emitting no carbon dioxide or other greenhouse gases into the atmosphere (though *all* power plants emit heat). If the industrial countries actually undertake the costly process of cutting back significantly on fossil fuel use as a precaution against global warming, as prescribed by the Kyoto protocol, nuclear power is the only electricity source capable of filling the resulting large gap in electricity supply. Renewable electricity sources such as solar, wind, and biomass could make a small and useful contribution but could not fill the gap alone (see Chapter 9).

Several important barriers stand in the way of a resurgence of nuclear power. The cost of building new plants is one of the most important barriers. While the capital cost of constructing nuclear power plants (about $2,000 per kilowatt) is considerably less than that of comparably sized solar or wind plants, it has not been low enough to compete effectively with coal (about $1,200 per kilowatt) or the latest generation of natural-gas technology (combined-cycle technology), which offers the advantages of lower construction costs (about $500 per kilowatt), higher operating efficiency, and competitive fuel prices. A further advantage of constructing natural gas plants is their low up-front costs, about 25 percent as compared with 60–75 percent for nuclear plants. But the fuel cost advantages of natural gas have recently eroded. Tight supplies and growing demand have caused U.S. natural gas prices to become very volatile in recent years, with fourfold increases in most locations and up to tenfold increases in California. Should gas prices continue to rise or should climate concerns bring about imposition of heavy taxes on carbon dioxide emissions from fossil fuels, new nuclear power plants could become competitive in the baseload electricity market. The situation with *existing* nuclear plants is more favorable, because their capital costs are largely amortized and their operation, maintenance, and fuel costs are relatively low. This is why the nuclear industry shows such a keen interest in renewing nuclear-plant operating licenses.[12]

NUCLEAR WASTE

A major barrier to the resurgence of nuclear power is the waste issue—the challenge of safe storage of spent fuel materials from plant operation. The acronym NIMBY—"not in my back yard"—aptly characterizes the prevailing public attitude toward spent-fuel storage. Because this remains a contentious political issue, radioactive wastes from decades of plant operations remain in nonpermanent surface storage at plant sites around the world, thirty-one in the United States alone. Yet the safety of such surface storage is much less assured (especially with regard to terrorist attack) than long-term storage in the underground burial sites that have been proposed.

Because of both political and technical issues, the number of nuclear waste repositories around the world will probably never be large. The U.S. Department of Energy has concentrated its recent efforts on a single geologic repository at Yucca Mountain, Nevada, intended for permanent storage of the wastes from the hundred or so nuclear power plants now operating in the United States. Extensive research has demonstrated that the very dry Yucca site and the repository design are compatible with safe long-term storage. Yet uncertainties remain; for example, the possibility that, over many centuries, groundwater might become contaminated by moving through tiny rock fractures and seeping into the storage site. And there are inherent uncertainties simply from the fact that repository data acquired over only a decade or so must be extrapolated for thousands of years.[13]

An important safety requirement for the Yucca Mountain repository is the removal of heat generated by the radioactive waste materials, which can be done by ventilation of the storage areas or by increased spacing of the waste canisters. The heat load will probably determine the repository's ultimate storage capacity. Also, the long-term potential for earthquake damage to the repository is being studied by searching for geological faulting at the site.[14] The resolution of such technical issues will largely determine the outcome of the Department of Energy's application to the Nuclear

Regulatory Commission (NRC) for a license to construct the Yucca Mountain repository by the end of this decade.

Technical issues aside, the main barrier to proceeding with a federal facility for permanent spent-fuel storage is political opposition, which remains very strong in Nevada, where the issue has assumed the dimensions of a states' rights conflict between the state and the federal government. Even if the NRC grants a license for Yucca Mountain on technical merit, the final decision on this controversial project may rest with Congress, which has the power to override state interests, and ultimately with the courts. In view of this clouded atmosphere, several decades may elapse before the nuclear-waste storage issue is resolved both technically and politically.

SPENT-FUEL REPROCESSING

It is technically feasible, though economically questionable, to chemically reprocess spent-fuel elements from nuclear power plants to separate out pure plutonium for use as fresh reactor fuel. Nuclear-waste repositories could also be designed so that at some future time, when their radiation levels have greatly decreased, the spent fuel elements could be reprocessed. Such reprocessing of spent fuel as a resource-conserving measure is not economically justified at present, but it could become so in the future as the cheapest uranium reserves are drawn down. Physicist John Holdren points out, however, that uranium would have to become about ten times more expensive than it is today for a reprocessing cycle to compete with a once-through cycle using low-enriched uranium fuel.[15]

Another argument against spent-fuel reprocessing is that the pure plutonium so produced could be diverted, with some effort, into the production of clandestine nuclear bombs. In fact, Britain, France, Russia, Japan, and India presently carry out reprocessing and among them separate enough pure plutonium from spent fuel each year to make over two thousand nuclear explosives. Reprocessing and plutonium-fuel fabrication plants are difficult to safeguard, and separated plutonium is at risk of theft by proliferant states and subnational groups, as well as risk of diversion by its owners.[16] The risk of diversion can be reduced, though not eliminated, by international agreements to prevent commerce in plutonium and, of course, by ceasing the practice of reprocessing. At present the United States "does not engage in plutonium reprocessing for either nuclear power or nuclear explosives purposes."[17]

BREEDING

The "breeder" is a type of nuclear reactor design that greatly increases the amount of nuclear fuel that can be obtained from natural uranium. Whereas ordinary reactors make use of less than 1 percent of the uranium (^{235}U) as fuel, breeder reactors convert the other 99-plus percent (^{238}U) into plutonium, also a nuclear fuel. If this type of reactor could be developed into a safe and economically competitive commercial system, it would indeed guarantee the availability of nuclear fuel for thousands of years, as mentioned earlier in this chapter. In recent decades the United States, United Kingdom, France, Japan, and Russia had large breeder-reactor R&D programs; but the U.S. and UK programs were terminated in 1994, Japan's breeder was shut down in 1995, and France's in 1998. Not only were cost projections for breeder technology unduly high, but technical and safety problems plagued all the programs. Even more important, however, is that the plutonium produced by breeder reactors is directly usable as nuclear weapons material. The potential for diversion of such plutonium into the manufacture of clandestine nuclear weapons is considered a serious issue, although technical fixes may eventually be developed. In any case, near-term commercialization of breeder technology is highly unlikely, as there will probably not be any economic or resource justification for breeder systems over at least the next several decades.

THE NUCLEAR FUTURE

When all factors are considered, the long-term prognosis for nuclear power is likely to depend most heavily on the cumulative safety record of nuclear power plants, as this will be a key determinant of public confidence in nuclear power. As of the present, the record is excellent. Over four decades of nuclear plant operation in the United States, not a single documented fatality involving radiation from nuclear plant accidents or waste materials has occurred, while thousands of fatalities have resulted from accidents related to other energy sources. A 1992 review of nuclear power in the United States by the National Academy of Sciences concluded:

- The risk to the health of the public from the operation of current reactors in the United States is very small. In this fundamental sense, current reactors are safe.

- A significant segment of the public has a different perception and also believes that the level of safety can and should be increased.

· As a result of operating experience, improved operator and maintenance training programs, safety research, better inspections, and productive use of probabilistic risk analysis, safety is continually improved. In many cases these improvements are closely linked to improvements in simplicity, reliability, and economy.[18]

Heralding a possible resurgence of public interest in nuclear power, a new family of uranium-fueled generating systems is being developed that will raise the safety level of nuclear power plants even further, and will operate with extremely low levels of risk to the public.[19] The safety systems of these new designs are passively controlled through the systems' basic physical properties and require neither operator intervention nor computer takeover in case of serious malfunction.

In keeping with the themes of this book, it is appropriate to remark on the potential for nuclear-power technology to contribute to the growth of electricity systems in the developing countries. Probably most relevant is the unusual complexity of nuclear power systems, whose construction requires heavy up-front capital investment and whose safe and efficient operation requires sophisticated and highly monitored infrastructures. With respect to the latter, the nuclear power industry is much like the air transportation industry, although it could be argued that the safety record of nuclear power in the industrial countries is considerably better than that of the airlines. In the fast-growing economies of Southeast Asia the capital and infrastructure requirements for nuclear power are in some cases already in place, consequently several nuclear plants are now on order or under construction in that region. In contrast, the world's poorest countries are not likely to be in a position to utilize nuclear power until their capital structures are more robust and their technical and administrative infrastructures more mature. China is an interesting case. In spite of the country's heavy endowment of coal resources, China is committed to building a number of nuclear plants, probably to ensure its future technical capability in this area. Yet coal will almost certainly remain China's fuel of choice for electricity generation. Generally, most developing countries will continue to rely on fossil fuels, especially coal, for their electricity generation, even though the environmental impacts from conventional coal technologies exceed those from nuclear power.

11

WHEELS

In no other area of human activity are the world of the rich and the world of the poor more disparate. In the world of the poor, the subsistence farmer travels on foot to barter his produce or fetch supplies for his family, covering but a few miles in a day. That same day, in the world of the affluent, the tourist or businessperson comfortably jets across half the globe.

Poverty and immobility, affluence and mobility—these invariably go together. Mobility is both a prime goal and a reliable indicator of development. During U.S. industrialization, rail technology boosted mobility and opened up new opportunities in a vast land. After World War II, the industrial countries achieved an unprecedented level of mobility as the personal automobile became its ubiquitous symbol. The new transportation systems allowed people to commute to work, deliver goods and services, widen housing opportunities, enjoy leisure travel, and expand the economy. During recent decades, many developing countries have been emulating rich mobility, and in all but the poorest countries numbers of automobiles have been burgeoning.

As with all other manifestations of development, transportation systems produce impacts on the environment, some beneficial and some harmful. These have appeared in both industrial and developing countries. Sometimes we forget that the environmental impacts of the automobile in early-twentieth-century American cities were very beneficial, as thousands of tons of foul-smelling, disease-ridden animal excrements gradually vanished from the streets. But with the huge and rapid growth in vehicle numbers and their urban concentrations after World War II, negative impacts such as air pollution and traffic congestion came to the fore. In the affluent countries, air pollution has been abated with increasing success by determined and continuing efforts over the last three decades, while in the developing countries the battle against air pollution has barely begun. (See Chapter 7 for discussion of air quality.) In contrast to air pollution, traffic congestion has only gotten worse and remains a major challenge in all vehicle-dominated cities, developed and developing alike.

In the industrialized countries, over five hundred million automobiles and trucks are now on the road. That's close to seven hundred vehicles for every one thousand people. In these countries, especially the United States and Canada, vehicle ownership is approaching saturation. Although affluent consumers may own several special-purpose vehicles, it is unlikely that the number of vehicles on the roads will grow much in the future. Not so in the developing world, where rapid growth in the vehicle population is just getting underway. The developing world now has about five billion people and about one hundred fifty million vehicles—an average population of only about thirty vehicles for every one thousand people. (In one of the poorest countries, Bangladesh, there are only 1.1 vehicles per thousand people.[1]) Were Thomas Malthus alive today, he might project a scenario in which the developing world's vehicle population increases to the point that the proportion of vehicles to people reaches the level of today's affluent countries. In that scenario there would be over six billion vehicles on the world's roads! Imagine, if you can, the megacities of the developing world even more crammed than they are today with antiquated, exhaust-belching cars, trucks, and buses, and you have an eye-burning, frustrating scenario that would make anyone cringe.

Fortunately, the world's vehicle future is not likely to resemble such a Malthusian debacle. For one thing, there is no way that the number of vehicles could reach anywhere near six billion in the foreseeable future. One recent study projects that the global vehicle total will reach 1.1 billion in 2020—an increase about equal to the industrial countries' total 1996 vehicle fleet.[2] A considerable increase and a congestion headache, to be sure, but probably manageable on the global level. Secondly, technological innovation is on the verge of spawning a radically new generation of vehicles that will be not only much more resource efficient than today's but also virtually pollution free. Early versions of these supervehicles are already entering the market in the affluent countries, and within another decade they will probably be appearing in the developing countries as well. Such vehicles have the potential to totally transform the world's urban environments during the next half-century.

THE NEW GENERATION OF SUPER VEHICLES

Although increasing numbers of vehicles will inevitably be on the road in coming decades as more and more people acquire the means to satisfy their desire for personal transportation, it is not at all inevitable that the physical environment—in cities, regions, or the globe—will suffer from the exercise of this freedom. A twenty-first-century high-technology vehicle revolution is quietly underway, promising vehicles so clean running that *the total pollution worldwide from a high-tech vehicle fleet twice the size of today's could be significantly lower than the pollution from today's fleet.* In the industrial countries it is likely that by mid-century the new high-tech vehicles will have largely supplanted the current generation of vehicles powered by internal combustion engines. Although advanced vehicles will penetrate the developing countries more slowly, they will play an even more crucial environmental role in rejuvenating the megacities now being ravaged by vehicle pollution, traffic congestion, and urban blight.

The new generation of vehicles combines the best characteristics of traditional, gasoline-powered vehicles with those of purely battery-powered vehicles—hence the name *hybrid.* Basically, the hybrid is an electric car that also has a small internal-combustion engine on board to charge its batteries while it is being driven. Because its batteries don't run down quickly, the hybrid overcomes the major limitation that has hindered consumer acceptance of the purely battery-powered electric car, namely, short driving range between battery charges (which can take several hours or overnight). In contrast, the hybrid's batteries can be charged continuously while driving; that is, until the small gasoline engine needs refueling (which takes only minutes at any filling station). The dual power system of the hybrid does add some complication to the vehicle's design, yet the technologies involved (with the exception of the battery systems) are mostly straightforward and well-known. Because few novel principles are involved, the hybrid development could be considered evolutionary. Yet the impact of hybrid vehicles on the global environment and the world's use of resources could be revolutionary.[3]

Hybrid technology is evolving rapidly, and hybrid vehicles will begin to join the world's fleet of conventional automobiles during this decade. In 2002 two major automobile manufacturers, Toyota and Honda, were already offering small hybrid models in the U.S. market, and the "Big Three" manufacturers are expected to have offerings by 2005. These models are harbingers of future vehicles that will offer both very high fuel efficiency and very low pollution, along with comfort and safety levels comparable to today's compact cars. As of this writing, early production hybrid vehicles are priced considerably higher than their subcompact counterparts with conventional engines, but there are no fundamental technical reasons why the costs should not become competitive with today's vehicles. Further, consumer acceptance in the environment-conscious affluent countries is expected to be high, and this will lead to increased production, lower unit costs, and lower prices, probably within this decade.[4]

The hybrid's main objective is achievement of high fuel efficiency and very low air pollution in a single automotive package similar in power, comfort, and price to today's vehicles. If the ambitious goals set for the recent R&D partnership between the U.S. government and several auto manufacturers are achieved, a hybrid automobile will emerge with a fuel efficiency of eighty miles per gallon, about three times the efficiency of today's models, and pollution emissions about one-eighth that of today's models. These goals are not just an engineer's dream—they are actually achievable. Allowing for the inevitable slippages in a technology development of this magnitude, one can confidently project that practical hybrid cars from the large manufacturers will be on the world's roads in significant numbers by the end of this decade. From that point on, growing market penetration around the world will depend upon several factors, including government regulatory and taxing policies and, of course, the progress toward affluence made by the developing countries.

Although the hybrid gasoline-electric automobile is the concept most nearly ready for the consumer market, it is by no means the only advanced technology that can lead to high fuel efficiency and very low pollution. The technology with the greatest long-term potential is probably the *fuel cell*–powered electric vehicle. This type of vehicle is driven by an electric motor that derives its electricity not from conventional batteries but from an onboard fuel cell. A fuel cell is a kind of battery that generates electricity from refillable gaseous or liquid fuels (such as hydrogen or methanol) rather than from the nonrenewable solid electrodes of conventional batteries. Since most fuel-cell vehicles are propelled only by electric motors, which emit no pollutants, they are truly "zero pollution" vehicles, in contrast to the hybrids, which emit tiny amounts of pollutants from their small internal-combustion engines.

Fuel-cell vehicles are the subject of intense and costly R&D programs being carried out by many automobile manufacturers, some with government assistance, and also by a number of independent firms working exclusively on fuel cell development. At the time of this writing, the experimental fuel cells themselves are too large and heavy to be practical for use in automobiles, but one manufacturer, Daimler-Chrysler, plans to build a number of fuel cell–powered city buses by 2003. When the inevitable market competition develops between the fuel cell and the hybrid, not only will consumers be the obvious beneficiaries, but the worldwide deterioration of air quality that accompanied the rise of the automobile culture will be permanently reversed, and the world's dependence on petroleum will probably be drastically reduced, as well.

Earlier in this chapter it was asserted that the total pollution worldwide from a high-tech vehicle fleet twice the size of today's could be significantly lower than the pollution from today's fleet. That this is a conservative projection is seen from the fact that today's first-generation hybrid vehicles emit only about one-eighth as much pollution as today's average car conforming to current U.S. environmental regulations. Even assuming that the emission levels of heavier, luxury hybrids joining the fleet are double that of other hybrids, we still wind up with much lower air pollution than today's. And, as stated, fuel cell–powered vehicles will emit essentially no pollution. Whichever technology—hybrid or fuel cell—ultimately wins out in the marketplace, the biggest winner will be the environment, since both types of vehicles are much cleaner running than today's internal combustion vehicles.

The world's love affair with the automobile shows no signs of abating, but the object of that love will be quite different in the future. One day, probably before the middle of this century, the world will no longer produce gas-guzzling, air-polluting vehicles, and in the affluent countries they probably will not even be tolerated. Yet the new vehicles will offer at least as much power, comfort, safety, and glamour as today's offerings. The technological advances underlying this remarkable transition are being generated through huge private and public investments in R&D, largely in the industrial countries, and by myriad environmentally oriented consumer choices in both developed and developing countries—choices made possible by the growth of freedom and affluence.

URBAN ROAD CONGESTION

No matter how clean running and resource efficient road vehicles may become, they will still take up space. Even the world's most efficient vehicles will not move people and goods if they are stuck in roadway gridlock. Drivers' time and patience have limits. In many U.S. cities, people already spend more than one hour a day commuting, and on major roads the peak morning and evening traffic flow can often be measured in inches per minute. Data from sixty-eight U.S. urban areas show that the average delay per driver increased 181 percent between 1982 and 1997 and 29 percent between 1992 and 1997.[5] Besides the frustration it provokes, gridlock is costly. In U.S. urban areas, lost working time and wasteful fuel use alone were estimated to cost $43 billion in 1990.[6] Actual costs, which also include increased air pollution and lower business productivity from delays in goods delivery, are probably higher.

People have grown weary of road congestion, just as they became weary of air pollution, and most believe that congestion is getting worse.[7] But in contrast to air pollution, effective solutions have thus far been elusive. In countries whose current living patterns evolved around the automobile, such as the United State and Canada, suggestions that people consider abandoning the automobile lifestyle are unrealistic. Strong constituencies have developed around other approaches to the congestion problem, including: (1) expanding the urban road network, (2) building more fixed-rail transit systems, (3) installing more bus systems, (4) using information or automation systems for cars and/or roads, and (5) employing various financial incentives such as congestion taxes.

MORE ROADS FOR MORE CARS

Throughout the world the personal automobile is the universal symbol of independence and affluence and a tough competitor to any other form of urban transportation. For convenience, comfort, and versatility, and for its unique ability to provide door-to-door transport without line changes and transfers, the car cannot be surpassed. In the United States, vehicle use continues to increase: between 1975 and 1998 the number of vehicle-miles of travel doubled, and the major U.S. road system expanded to its present length of 3,966,485 miles (2002 data).[8] Yet, despite continuing road construction, the highway network is increasingly stressed, especially in and around urban areas, and more than half of peak-hour traffic in urban areas occurs under congested conditions.

For most cities, building new roads is less likely to be considered a viable solution than it was two decades ago. Experience has shown that, as a rule, adding new roads simply attracts more vehicles. Some commuters will switch to a new road from other roads; others will abandon public transit and return to driving; still others will stop ride sharing and drive alone. In short order the road system becomes every bit as congested as it was before expansion. In the United States, some urban road building still goes on—for example, in Boston, where the Central Artery/Tunnel construction is part of a huge urban redevelopment project; however, the relief of traffic congestion will probably be only temporary. In contrast, in San Francisco several urban roadways damaged by a 1989 earthquake were actually removed rather than rebuilt. Throughout the country, current efforts are concentrating on increasing the use efficiency of the present road system. (See below.)

PUBLIC TRANSIT

Public transit systems are clearly an urban necessity, because many citizens are too poor, too young, too old, or too ill to drive. But what kinds of systems best serve the public, and who should pay for them? Before World War II the industrial countries relied heavily on rail-based transit systems as the principal urban people movers. These systems were designed to fit the relatively compact housing and employment patterns of cities that matured in the early part of the twentieth century, before the spread of the automobile culture—cities such as London, New York, Paris, Chicago. Today, rail-based transit systems continue to serve the central city's markets adequately, but they have not been able to halt the increase in road congestion that accompanies the continuing growth in suburban sprawl and automobile commuting. Cities that grew to maturity *after* the automobile revolution, such as Los Angeles, Phoenix, and Dallas, are laid out with a dispersed geography that is poorly suited to fixed-rail designs. (Unfortunately their road systems have also proved incapable of handling the continually growing automobile traffic.) Rail-based public transit systems continue to have strong advocates, especially among environmentalists and urban planners. Several new rail systems built in the 1970s and 1980s (San Francisco, Washington, D.C., San Diego) have been well received, but only the very costly D.C. system, paid for by a national subsidy, is comparable in its spatial extent and impact to the venerable New York and Chicago systems.

The five thousand–plus public transit systems in the United States have suffered from two chronic problems: insufficient ridership and insufficient funding. The main causes of poor ridership on central-city rail systems are (1) ongoing suburbanization and increased auto use and (2) existence of competing suburban transit systems that better serve the growing number of commutes from suburb to suburb rather than from suburb to city (the traditional pattern of the large existing rail systems). Between 1984 and 1995 U.S. transit systems lost almost 15 percent of their total ridership, with most of the loss coming from the largest urban systems.[9] Since then, probably influenced by increasing traffic congestion and commuting delays, transit ridership has picked up again, and the loss was recovered by 1999. This reversal has led to increased optimism among transportation professionals that public transit will be able to contribute significantly to meeting communities' mobility needs in the coming decades.

The recent economic history of transit systems, however, provides meager basis for optimism about the future of publicly funded urban systems. Even the most efficient of existing systems normally recovers less than half of its operating costs from rider fares (deliberately set low so as not to discourage patronage). Operating costs have been rising rapidly, and most of the costs must be subsidized either by government or by local taxpayers. Between 1988 and 1998 federal, state, and local investments in U.S. transit systems nearly doubled, from $3.8 billion to $7.1 billion, and state and local governments used $4.9 billion of federal highway funds for transit systems.[10] Yet the history of cost overruns, inadequate cost recovery, and inefficient management of publicly funded transit systems continues to fuel opposition to subsidies among those who believe that privately funded systems would better serve the public's transportation needs.

Some critics of light-rail systems favor expanded use of bus lines, which can arguably do the job better and for less money. Buses have the advantage over fixed rail of being much more flexible in routing and better able to accommodate time-of-day load variations. Disadvantages are that buses do not decrease road congestion significantly unless dedicated lanes are provided for them; they are also typically less comfortable than trains and are commonly perceived as a lower-class travel mode. Yet there are conspicuous bus-success stories: for example, the car-loving city of Houston has successfully reduced roadway congestion by constructing seventy-one miles of special bus/carpool lanes financed with a 1-cent sales tax.

INFORMATION SYSTEMS AND VEHICLE AUTOMATION

Some relief to road congestion may be on the way in the form of computer-assisted vehicle control systems (the auto equivalent of the aircraft autopilot), which can greatly increase the efficiency of vehicles' use of road space by allowing tighter car-to-car separation to be safely maintained. Although hopes are high in the industry for vehicle automation technology, it is much too early to tell when and by how much this type of technology will ease congestion and increase driving safety.

Another approach to more efficient use of roadways is the development of computerized information systems that inform both drivers and road management personnel how each section of roadway is performing in terms of traffic speed and flow. One prototype technology involves motion sensors embedded at frequent intervals in the road, transmitting information to central computers that can be accessed randomly from individual vehicles. Before embarking on a commute, a driver could obtain a computation of the fastest trip time for several alternative routes. For example, the driver would query, "tell me the best route if I start fifteen minutes from now." The result would be more efficient use of the road system by spreading out the commute peak and reducing the total time delay experienced by all commuters. Such information can also help road-management personnel achieve optimal control of ramp metering rates and advisory messages.[11]

EMPLOYING FINANCIAL INCENTIVES

It's called "demand management" by economists, and in the context of transportation it is simply a strategy to change travel behavior by using variable toll pricing. This approach is politically controversial, to be sure, yet variable pricing is a potentially effective policy for reducing gridlock. Congestion pricing, one form of demand management, would impose toll fees on public roads that vary depending on the degree of congestion on particular road segments at particular times. Drivers would not need to have coins available nor be required to slow down, because the fee would be automatically charged to the driver's account as the car passes each check point.

Here's an example of how congestion pricing might work. Suppose that a 25-cent toll were charged on a given road segment at its lowest traffic flow point (say, 3 AM) and that the toll of that segment were progressively increased to a maximum of 3 dollars at the peak commuter times (say, 7–9 AM and 4–6 PM). The typical cost for passing through that segment would be 3 dollars at peak hours, 1 dollar during the hour before and hour after the peak, 50 cents at other hours, and 25 cents during the late-night hours. Since many motorists would respond to the highly variable fees by adjusting their travel schedules, traffic on the segment would adjust until it reaches a more even distribution. The specific time-of-day fees would be adjusted frequently by computer analysis so as to keep the roadway congestion at a minimum for the total traffic volume.

Without the new generation of electronic toll-collection devices and computerized traffic-measuring sensors, the congestion-pricing strategy would be quite difficult to implement. But now that it is becoming technologically feasible, the major issue becomes public acceptance of the idea. Its technical attractiveness notwithstanding, congestion pricing is very difficult to sell politically, and, as of this writing, not one U.S. road system has adopted such a scheme. Opposition is based on the regressive nature of the road tax, which in its simplest form discriminates against low-income drivers and those who have no choice but to travel during peak hours. It is possible that this disadvantage could be mitigated by automated subsidies, but that approach would introduce other issues, such as fairness, as well as potentially severe administrative complications.

THE BOTTOM LINE ON ROAD CONGESTION

Road congestion has been one of the most pernicious environmental impacts of the automobile culture in the industrialized countries. Congestion is widespread not only in North America but also in Europe, where the number of vehicle miles traveled per mile of roadway is greater in Italy, Britain, and Germany than in the United States. Although road congestion is understood to be a reversible problem, citizens of all affluent countries are increasingly bothered by the deterioration in quality of life it causes, and they are impatient for solutions. Are there grounds for optimism that congestion will be relieved any time soon by some combination of the approaches just described? Two considerations point to the solvability of this problem: first and most important, the number of vehicles on the road is approaching saturation in the affluent countries. Second, affluent countries have the financial means to implement solutions if difficult political issues are faced and all reasonable options explored. As with many other environmental issues, affluence is a major key to the solution.

TRANSPORTATION IN THE RAPIDLY DEVELOPING COUNTRIES

How the advanced developing countries work out ways to transport people and goods in their teeming megacities will largely determine the quality of their citizens' lives in this century. Visitors to the great cities of the developing world quickly become aware that, with few exceptions, their rapid growth is extracting a high price in terms of environmental degradation and other problems of mega-urbanization. As bad as road congestion appears to motorists in the industrial countries, it is typically much worse in the developing countries. And progress in curbing air pollution lags far behind, as well; in many cases there has been no progress at all. For example, the notorious traffic congestion, noxious pollution, and high accident rates on the roads of Bangkok impose costs as high as a billion dollars a year and reveal the shortcomings of urban transport planning as one of the most formidable challenges attending the transition from poverty to affluence.[12] Some cities in developing countries, including Bangkok, Calcutta, Seoul, Mexico City, Tehran, and Buenos Aires, are already more congested than any city in Western Europe, even though car ownership levels are only a third as high.[13]

In Asian cities every conceivable mode of urban transport is being used—bicycles, pedicabs, rickshaws, motorcycles, private automobiles, taxis, buses, articulated trolleys, monorails, rail transit systems—and walking. In spite of this modal variety, rapid urban growth and massive in-migration have been accompanied by deterioration in transport services, which often suffer from poor planning, excessive regulation, inadequate financing, and lax maintenance. Many cities have neither increased nor improved their woefully inadequate road space, and roads often do not exist in the poorest areas. The needs for pedestrian areas, mostly used by the poor, are usually sacrificed to the needs of vehicles, mostly used by the rich. In many developing countries a heavy environmental price is being extracted by the collective mishandling of transport.

Although the fast pace of economic development may exacerbate the difficulties of urban transport in some countries, affluence itself is certainly not the root cause of the problem. Of all the Asian countries, the most affluent—Japan and the city-states of Singapore and Hong Kong—have also been the most successful in developing more-than-satisfactory urban transport systems. For example, of the mass-transit systems recently built in twenty-six large Asian cities and already carrying seventeen million passengers per day, over half are in affluent Japan, which also has 70 percent of the region's automobiles.[14] Affluence is an important ingredient of transport solutions, yet it cannot substitute for competent urban planning and democratic decision making, ingredients that have often been lacking in developing countries.

TECHNOLOGY FOR REJUVENATING CITIES

Because people's universal appetite for personal vehicles is not likely to wane, approaches to rejuvenating the megacities of the developing world must accommodate the car's ubiquity. Fortunately, the new generation of environmentally superior vehicles has the potential to produce a sea change in urban environmental quality. Several of the developing countries, among them China, India, and Brazil, are large enough to influence the commercial evolution of these technologies so that a portion of the vehicles marketed worldwide meet the specific needs of developing economies, for example, availability of inexpensive minivehicles for first-time buyers. Government incentives will undoubtedly be needed to encourage the replacement of old gas-guzzling polluters with fuel-efficient, low-emission vehicles as these become available.

Although in the long term advancing technology will relieve the developing countries from having to grapple with most vehicle-generated air pollution, over the next two decades it is essential that they emulate the most successful of the pollution measures developed by the affluent countries, including regulating and monitoring vehicle emission levels, using cleaner fuels, and encouraging fleet turnover, especially of old diesel trucks and buses. Also in the near term, it is critical that they unplug the time- and money-wasting traffic congestion that is especially severe on developing country urban roadways. Although urban transport systems and infrastructures differ widely among countries, some common problem and solution areas can be identified.[15] First, adequate public transit needs to be provided, with lines that go where people live and where they work. Transit systems should be accessible, dependable, and well maintained and should be managed and operated by qualified personnel. Even with heavy rider utilization, subsidies will be necessary to cover costs, but users should be required to carry as much of the cost burden as politically feasible.

Second, effective demand management is required to reduce traffic congestion. Separating cycle and pedestrian traffic from motorized traffic has rarely been done in developing countries, but it is critical for reducing congestion. Fast lanes reserved for buses and cars with multiple riders are also necessary. Most important, more and better roads are needed, and road users should pay the full costs of building and maintaining them through vehicle fees, roadway tolls, fuel taxes, and parking charges. (Very few countries in the world, including the affluent ones, now require full cost recovery for road systems.) Restrictive driving regulations, such as odd/even license number days or restricted traffic zones, should be adopted only with caution, because such schemes can be counterproductive as drivers find ingenious ways to bypass the rules, for example, by purchasing additional cars or driving longer distances to reach destinations.

TRANSPORT TO AFFLUENCE

Just as in the past, transport systems today serve as powerful catalysts for economic and social development. Crucial to the growing global economy, modern transport systems will play a vital role in integrating new networks of communication and trade with those that have long flourished among the affluent countries. Transport systems are complex and functionally interdependent, requiring not only costly infrastructure but also intricate coordination among various air, sea, and land modes. The complicated environmental problems associated with these rapidly evolving systems also take considerable time and effort to mitigate.

One of the best examples of transport progress is Singapore, which since 1960 has developed solutions to its intense traffic problems in the larger context of supplying housing, jobs, and income security for its people. New transportation infrastructure became the basis of Singapore's urban redevelopment program and allowed planned communities to be created on the outskirts to accommodate industrial and population growth. Scenic boulevards and waterfront parkways made possible the transformation of old city slums into desirable new housing units as well as industrial areas, parks, and schools.[16] Although not all developing countries enjoy the possibilities for transport that compact Singapore has, Singapore's unique experience does demonstrate how important transport can be in achieving goals of societal development, including jobs, education, and industry growth, which in turn catalyze further development.

TRANSPORT FOR THE POOREST OF THE POOR

In the world's poorest countries, many located in sub-Saharan Africa, the transport "system" consists of a few narrow dirt roads full of ruts and gullies—sun-baked generators of dust when dry and impassable streams when rain soaked. The modes of transport are the feet or the draft animal: a world not of this century but of another age—a world apart from the megacities of the developing countries.

There is rich history and vibrant human life in these areas—families, farms, villages, schools. But lack of mobility severely limits access to resources and commerce, adoption of modern farming practices, delivery of education and medical care. People still living in isolation must be able to communicate and interact more readily with their fellows and with the outside world. Many different approaches are needed to address the multiple causes of poverty. Transportation's role is to overcome the immobility and inaccessibility that prevent other elements of the problem from being addressed.[17] Few needs are greater than the need to possess elemental systems of all-weather roads and transport services connecting villages and towns, schools and markets. One of the most challenging and potentially rewarding targets of international development aid is the establishment of basic road and transport infrastructures wherever local progress toward stability and freedom warrants. The poorest countries need not only construction capital and expertise but also continuing support for upkeep and maintenance. Though a chasm of development separates the small inter-village bus line from the huge urban transit system, for many riders the little bus may offer the first ride on the long journey to a better life.

12

DON'T HARM THE PATIENT

The oath of Hippocrates is usually considered the most fundamental ethical guide in the practice of Western medicine. It states, in part: "And I will use regimens for the benefit of the ill in accordance with my ability and my judgment, but from what is to their harm or injustice I will keep them."[1] Simply put, help the patient if you can, but above all *don't harm* the patient.

Have human activities irreparably harmed patient earth?[2] Although this question elicits strong yeas and nays from various quarters, we really don't know the answer. What we do know is that, ever since life began, all species that roamed the earth have altered the environments in which they lived. And we humans have affected our environments far more than other species. The development of modern industrial societies left, along with many benefits for human life, a huge trail of environmental damage, especially air and water pollution. Fortunately these impacts were largely reversible, and throughout these pages I have noted the affluent societies' strong efforts to restore and protect their environments, mostly with success.

But not all environmental impacts are as reversible as air and water pollution. If a particular species of plant or animal becomes extinct throughout the planet, it cannot be retrieved, it is gone forever. The incredible variety of the earth's life forms—the earth's *biodiversity*—is thereby reduced. It is no wonder, then, that the subject of biodiversity has become so important to biologists. But biodiversity should also be important to the rest of us, for it is the totality of plants and animals, and the seamless webs of their interactions, that constitute Nature, which sustains and enriches human life on earth. How are human activities affecting the diversity of plants and animals? If species are being lost, at what rate are they being lost? How important is species loss? What do reductions in biodiversity portend for the earth's future? Are the affluent countries doing enough to protect the species with which all of us, affluent and poor, share the planet?

Despite the serious scientific dialogue about biodiversity, doomsday rhetoric abounds in the media, as, for example, the advertisement, cited in the Introduction, claiming that "entire ecosystems are in danger of disappearing forever . . . and the fate of the planet rests on choices we make today."[3] Although most people would have a hard time accepting that the fate of the planet is as precarious as the ad suggests, they would undoubtedly agree on the desirability of maintaining the planet's rich diversity of plant and animal species. But they would not necessarily share the same motivations for preserving species diversity. Here are four perspectives on biodiversity.

THE INTELLECTUAL PERSPECTIVE

Evolutionary biologists are interested in studying the millions of individual species and their ecological relationships because these provide a unique window on the history of evolution. Understanding these relationships can contribute to an understanding of not only the development of life but also the place of *human* life in the complex web of the earth's living things. One example: evolutionary biologists have studied a large group of related fish species (called Cichlidae) in the lakes of East Africa. This group of species has shown an unusually rapid rate of speciation; for instance, in Lake Victoria, over seven hundred species of cichlids have evolved in the past thirteen thousand years.[4] Many of these species are currently threatened, and biologists fear that this window on the processes of evolution may be in danger of closing irretrievably.[5]

THE ETHICAL PERSPECTIVE

Even more important than gaining the knowledge biodiversity holds, biologists believe, is protecting the evolutionary process itself, that is, maintaining a diverse gene pool for future evolution. Extinctions of species reduce the gene pool, and this decrease in genetic variability affects both the future potential for unique evolutionary events and the ability of species to survive. If such diversity is lost, it may never be replaced; the stores of scientific knowledge not yet gained might never be gained; the evolution that would have taken place might never take place. Fearing such losses, and aware that biodiversity's scientific value cannot be translated readily into monetary terms, biologists tend to harbor considerable pessimism about the future prospects for biodiversity.

Since the beginning of time, all species have altered their environments, yet Homo sapiens has a global reach that exceeds all other species and is uniquely capable of altering the course of evolution for millions of years in the future. Although such a consequence is not necessarily detrimental to the planet, many evolutionary biologists believe that the *possibility* of harm poses a deep ethical issue, first for the scientific community, then for the larger human community. Biologist Paul Ehrlich puts the issue this way:

> Revenue from logging a tropical forest might be used to help poor people living near the forest today. Would it be worth forgoing that revenue to preserve the forest as a potential generator of [biological] diversity that might improve the lives of people 2,000 or 200,000 generations in the future? How are values to be assigned, and who should make this sort of decision? Is there any ethical need to consider the effects of today's actions that far or farther in the future? Could or should we strive to create such an ethical imperative? Can we possibly know enough to sensibly fashion an evolutionary ethic?[6]

THE SPIRITUAL PERSPECTIVE

Whoever has stood on a mountaintop and beheld the vastness of land and sea or trekked through a fog-shrouded ancient redwood grove, whoever has experienced the wonder of a giant condor soaring above or a mother lion playing with her cubs in the wild understands the spiritual perspective. For many, such experiences of nature's grandeur are the essence of spirituality. For some, they bring the feeling of closeness to a supreme being in the same humbling way as the cathedrals Notre Dame or Chartres. Such a feeling motivated naturalist John Muir over a century ago as he worked tirelessly to secure vast tracts of undisturbed land for the public trust. And such a feeling motivates countless numbers of environmentalists around the world today. From the spiritual perspective, preserving nature's richness and beauty is a moral imperative to which humans should be committed. It has nothing to do with science.

THE ECONOMIC PERSPECTIVE

This perspective recognizes that the human species makes use of nonhuman species in a multitude of ways that have economic value, many of which remain to be discovered in the future. We use plant and animal species for shelter and warmth; we make food, clothing, and medicines from them; we admire them in zoos or jungles and keep them as pets. According to the economic perspective, species diversity should be conserved so that humans can continue to make use of other species in a sustainable way. This viewpoint is anathema to moral preservationists, who believe that nonhuman species have the same right to life as humans, and also to many scientists for whom the value of ecosystems is intrinsic. Most of the economic uses of species (and their products) involve trading in the marketplace, yet whether this is the case or not, the species tend to be undervalued because the costs of maintaining a sustainable supply are usually not included in the costs to either the provider or user. Such supply costs are neglected, for example, when native forests are cut back to provide fuelwood (whether free or commercial) or when animals are killed in the wild, either by rural folk who depend on them for food or by traders who sell their skins or tusks.

Each of these perspectives has validity in its own context. Scientists want species, whether lowly insects or charismatic jaguars, to be preserved because of their intrinsic value and value to future evolution. To economists, human welfare is the key; investments are justified to preserve species that can satisfy present and future human needs and wants. The spiritualist would maintain the grandeur of nature, if necessary even at the expense of human welfare. Of course these contexts are not always distinct; often they overlap and subsume other positions. Scientists, for example, sometimes reach for economic justifications to enhance their arguments for preventing species loss, as when they cite the immense economic potential that biodiversity provides for developing new medicines, crops, pharmaceuticals, timber, fibers, and pulp. And economists contend that there is economic value in the spiritual benefits of biodiversity, pointing to nonconsumptive uses that are becoming increasingly important— for example, ecotourism and bird-watching, one of the fastest-growing outdoor recreational activities in the United States. Though sometimes confusing, these "mixes" are reasonable.

It is treading in dangerous waters, however, when economists challenge ecologists to quantify accurately the current rate of species extinction, as a demonstration of the seriousness of the species extinction problem.[7] For it is never justified to assume that in science only those problems that can currently be quantified may be considered important. When experts' collective judgment, based on scientific intuition as well as hard data, points to the seriousness of a particular problem, such judgment should not be taken lightly. In fact, evolutionary biologists concede that science has not established how many species now exist (estimates run to over 30 million species, of which only 1.5 million have been named), let alone how many species have become extinct in the past or how fast they are becoming extinct today. Many biologists accept a rough estimate of 0.1 percent per year as the current species extinction rate, a figure that is conceivably a thousand times greater than in prehuman times.[8] Examining the same data, statistician Bjorn Lomborg argues that the current extinction rate for animals is much lower, more like 0.014 percent per year, "not a catastrophe but a problem."[9] Whichever number turns out to be more nearly correct, however, the problem remains. Although such estimates, based partly on models of habitat reduction and partly on indirect empirical evidence, are highly speculative, there is enough science behind them for biologists to make a credible case that species may be disappearing at a dangerous rate, one that could portend serious consequences for the earth's biological future unless humankind reverses the extinction trend.

Nonetheless, in a democratic society neither ecologists nor economists nor spiritualists hold a privileged position from which to dictate to society whether or how much it should invest to preserve individual species or their habitats. Such decisions properly belong to the political process, by which proposals from scientists and others to commit public resources to protect biodiversity can be scrutinized in comparison with other items on society's menu of environmental priorities. In the United States the idea of protecting individual species underwent such political scrutiny and in 1973 was enacted into law as the Endangered Species Act.

The Endangered Species Act (ESA) embodies many of this book's themes and can even be seen as one of its major focal points. It is probably the most far-reaching environmental statute ever adopted by any nation. The act is solidly grounded in the moral commitment of the American people to preserve their environment and is a demonstration of the claim, made throughout this book, that free and affluent people will take action to protect their environment when they perceive an important problem and believe there is an effective solution. Scientists, environmentalists, and legislators played important roles, to be sure, but at bottom the Endangered Species Act belongs to the American people. Such a mandate, involving huge expenditures of public and private money, could not have come out of a country whose citizens were not dedicated to environmental quality.

Nor could the act have come out of an impoverished country. In fact, the gap between rich and poor countries in biodiversity conservation investments is enormous. In the developed countries, the average investment in

ted areas is about $1,687 per square kilometer, whereas in the poor
ries the average investment is only $161. This despite the fact that
both the biological diversity and threats to that diversity in poor countries
are often much greater than in rich countries.[10]

Just what is the Endangered Species Act? Simply, ESA requires the
federal government to identify and publish lists of species that are in
imminent danger of becoming extinct (endangered) or likely to become
endangered in the future (threatened).[11] The act requires that decisions to
list species be based on biological factors alone, without consideration of
economic factors. Requests for listing may be made by anyone who supplies evidence. A recovery plan for each listed species must be developed in
which economic factors may be considered, and the recoveries must be
monitored by the relevant government agency. Exemptions to the ESA
may be granted when the benefits of a proposed action affecting a listed
species clearly outweigh the alternatives. As of early 1997, 1,067 plant and
animal species were listed as endangered or threatened, and 644 species had
approved recovery plans.[12] Some 4,000 other species are on "waiting lists."

The Endangered Species Act had an inauspicious beginning. Its first test
case was a conflict between an obscure fish and an unimportant dam that
was taken all the way to the U.S. Supreme Court. The act was successfully
invoked to stop construction of a nearly completed Tennessee Valley
Authority dam, on which $90 million of public funds had already been
spent, to protect a tiny minnow called the snail darter. In its ruling, the
Supreme Court pointed to the act's wording, which "shows clearly that
Congress viewed the value of endangered species as incalculable." An
economist critical of the ruling quipped, "Obviously, a $100 million dam
was worth less than an infinitely valuable fish."[13] Favoring the construction halt, the Sierra Club took the position that the dam "was a classic TVA
pork-barrel project, justified less by flood control and hydropower needs
than by the number of construction jobs it would bring to Tennessee."[14] In
the end, both sides were able to claim victory. An exemption was passed by
Congress (prodded, to be sure, by the Tennessee delegation) allowing the
dam to be completed "notwithstanding the Endangered Species Act or any
other law." And the snail darter was relocated to other rivers, where the little fish now thrives.

This case highlights the most difficult policy consequence that ESA has
faced throughout its three-decade history. Even though the Supreme Court
interpreted the act to mean that Congress considers the value of endangered species to be "incalculable," in fact Congress has never appropriated
funds commensurate with the enormous costs involved in carrying out the

act's provisions. One estimate posits these costs to be in the range of $7 billion to $13 billion,[15] and another that "conservative estimates of the Act's costs are in the tens of billions."[16] Yet the annual budget of the U.S. Fish and Wildlife Service's endangered species program is only $58 million. For one species alone, the grizzly bear, consider the costs of providing habitat to support a minimum viable population of two thousand grizzlies, estimated variously from thirty-two million to almost five hundred million acres. Even the lower estimate represents an area equal to one-third of Montana.[17]

Since the benefits of preserving species and their habitats accrue to society as a whole, it would seem reasonable that the public should collectively bear the costs of carrying out ESA's mandate. All too often, however, private individuals or firms have been asked to pay all the costs. In one case, the economic activity around the town of Bruneau, Idaho, was threatened when the Fish and Wildlife Service (FWS) began cutting off water rights to fifty-nine farms and ranches in order to protect a local snail. The water flowed to the farms again only after a federal judge removed the snail from the ESA's list, citing inadequate scientific data.

Another example: in April 2001 the livelihoods of fourteen hundred farm families in Oregon's Klamath Basin were placed in jeopardy when the Bureau of Reclamation, evoking the ESA during drought conditions, cut off the farmers' irrigation water from Klamath Lake, which had sustained two hundred thousand acres of cropland for nearly a century. Instead, the water is being used to protect two species of suckerfish that inhabit the lake, as well as the downstream Klamath River coho salmon. To complicate the matter, only the wild coho is on the endangered list, while the hatchery-raised coho is not protected and is in plentiful supply. The National Academy of Sciences has been asked to evaluate the scientific information that led to the government's decision to cut off the farmers' irrigation water. Meanwhile, the farmers are being provided with "drought relief" instead of water.

Conflicts between traditional private property rights and the legislated rights of endangered species have been at the heart of continuing debate and acrimony over the Endangered Species Act. About half of the listed endangered and threatened species are thought to be found on private land, and the Fish and Wildlife Service has reported to Congress that "approximately 25 percent of all listed species have conflicts with development projects or other forms of economic activity."[18] Private landowners who wish to work around an endangered species found on their land can face extra costs as projects are terminated or delayed. Sometimes the discovery of an endangered species on a privately held parcel will reduce the parcel's market value, because of costs of protecting the animal or plant and because of government restrictions placed on the land's future use, for example, prohibition of logging on the parcel. In one case, property values in Travis County, Texas, dropped $359 million after the golden-cheeked warbler and black-capped vireo were listed as endangered, and one property owner saw the appraised value of her land decrease from $830,000 to $38,000.[19]

An important legal issue is raised by decreases in property values brought about by government actions under the ESA. On one hand, landowners are increasingly construing such government actions as an infringement of property rights, and the property value losses as the equivalent of government's "taking" of private land, which is prohibited under the U.S. Constitution unless accompanied by just compensation. On the other hand, supporters of ESA generally oppose government compensation for lost value, claiming that compensation, or even the potential for compensation, would destroy the noneconomic basis of species listing. In this view, loss of value is akin to the economic consequences of zoning ordinance changes, which also can affect the uses of private land. Thus far, the courts have upheld the latter position, much to the ire of affected landowners. If the courts ever change their position or if Congress enacts legislation requiring federal agencies to compensate property owners for losses under ESA, the costs of protecting species under the act could, obviously, become much higher. But higher costs may be an inevitable outcome of the "incalculable" value that Congress placed on each protected species in the original legislation.

Some useful strategies have arisen for negotiating the interests of parties affected by the designation of endangered species. Among these is the development of habitat conservation plans, whose intent is to proactively address conflicts between habitat conservation and land development by persuading all parties to agree to a conservation plan. An important example from the early 1990s is the Multiple Species Conservation Plan for the San Diego area, which focuses on the coastal sage scrub habitat, rich in threatened species. The San Diego MSCP was approved in 1997, setting aside 172,000 acres of open-space conservation land and issuing incidental take permits in other areas where development would be permitted.

Through this approach equitable solutions have been found in a number of other cases as well. For example, the Fish and Wildlife Service delayed construction of a hotel complex and hundreds of homes planned on 121 acres of private land overlooking Dana Point Harbor in Southern

California after some pocket mice were discovered during an environmental survey of the project. The conflict was resolved by cooperation between community and developer, resulting in a plan that balanced the property owner's rights, the community's rights, and the ESA's mandate to protect the pocket mice. The scale of the planned hotel was cut back, the number of new homes was reduced, and sixty-two acres of public open space and thirteen acres of private open space were created, to the satisfaction of most residents and, presumably, the pocket mice as well.

Given these examples of problems and solutions, by what overall criteria should we judge whether ESA is succeeding or failing? Supporters point to the more than one thousand species listed since 1973, the recovery plans in place for about half those listed, and the "waiting list" of four thousand new candidates. Since the ultimate aim of the act is not only to list endangered species but to return them to healthy population levels, the best measure of success may be the number of recovered species that are removed from the list. But this may not be a speedy result, because even in favorable cases many years can be needed for species to recover. Yet in May 1998 the Department of the Interior did announce that over two dozen species would either be downgraded or removed from the lists, and this was enthusiastically cited by the Interior secretary as proof that the Endangered Species Act works. The "proof" was questionable, however, as five of the delisted species were already extinct, and at least eight others had been erroneously listed because of incorrect taxonomy or because their numbers were greater than originally believed. Opponents of the act cited this gaffe as evidence of the act's failure, and one outspoken critic stated in a congressional testimony, "not one [of the twenty-nine species removed] was the result of an actual recovery plan."[20]

Love it or hate it, the Endangered Species Act is here to stay because the majority of the American people believe in it. Yet both supporters and critics are well aware that its high moral purpose and good intentions did not of themselves produce good policy. The act has glaring weaknesses and needs reform and improvement. Some of the areas that need attention:

· IMPROVED SCIENCE. The criteria for listing species are weak; in some cases selections have been arbitrary, in others political. Since cost considerations will probably continue to be excluded from the listing process, scientific considerations should be strengthened. The act should require preparation of a solid scientific case for each species of plant and animal listed, including both existing and proposed listings. The scientific reviews should focus on the importance of candidate species to the health of the ecosystems of which they are a part and should not favor charismatic or highly publicized species. In the case of habitat conservation plans, more effort is needed to assure that they are backed by solid science, for example, reliable data bearing on the likelihood that chosen mitigation measures will succeed. Peer review should be integrated into the entire process, from rank ordering the listings to establishing priorities for recovery plans and designating critical habitats.

. ECONOMIC CONSIDERATIONS IN ESA ACTIONS. Despite wide support for retaining the "incalculable" status of species in the *listing* process, most observers concede that, in the planning of *protection* and *recovery* actions under the ESA, cost considerations are critical and should be given more consideration along with scientific factors. In the government's development of habitat conservation plans, early discussions with landowners would help avoid economic conflicts and greatly improve the chances that planned government actions will be politically acceptable as well as scientifically sound. It is equally important for planners of private land-use projects to meet with government at an early stage, especially when the projects involve land designated as a critical habitat of one or more listed species.

· INCENTIVES FOR SPECIES PROTECTION. All agree that voluntary programs can play an important role in encouraging citizens to be stewards of the natural environment. But individual landowners should not be asked to carry inequitable financial burdens in complying with the act in ways that benefit society as a whole, for example, foregoing commercial use of a private land parcel to help save an endangered bird. A variety of financial incentives should be incorporated into proposed recovery and habitat-conservation plans under the act in order to encourage private landowners to take specific land-use actions or refrain from others. These could take the form of tax incentives, restrictive easements, or outright government purchases of parcel segments. Private funds can also contribute, as, for example, a fund maintained by the organization Defenders of Wildlife that encourages private landowners to allow gray wolves to breed on their property and reimburses livestock owners for documented losses caused by wolves.[21] (Not surprisingly, most ranchers are less than enthusiastic about the possibility that wolves may return, since their forebears went to such lengths to exterminate them.[22]) In the main, however, the preservation of species should be seen as a public good, and the costs of such preservation should be borne principally by public funds. As the economic realities of the Endangered Species Act become increasingly clear, Congress should provide responsible agencies with funding levels that more closely match the actual costs of implementing the act.

PROTECTING BIODIVERSITY IN THE LONG TERM

Although basic scientific research is not directly connected to either the listing process or recovery actions under the Endangered Species Act, science continues to provide the act's main intellectual underpinning. Even this chapter's brief look at biodiversity shows that much of the conflict over ESA arises from the inadequacy of the knowledge base. Scientists are still a long way from being able to simulate individual species–habitat systems in enough detail to guide specific policy actions—for example, deciding how much a government agency should invest to protect a particular bird on a particular property. But it is not unrealistic to expect that habitat simulations, in combination with field data, will often be able to provide credible cost guidance on conservation actions at a more aggregate level. An example is recent work by biologist Edward O. Wilson focusing on the costs of protecting large expanses of key habitats. In one specific case, Wilson suggests that it would cost about $5 billion to buy out loggers in the tropical forests of Amazonia, Congo, and New Guinea.[23] This sum is large but probably not beyond the reach of the international conservation community.

Looking to the longer term, we must depend on basic science research to provide the knowledge base for improving the Endangered Species Act and developing additional legislation to protect biodiversity. Government at all levels should support experimental research in the relevant sciences, especially measurement and tracking of population sizes and dynamics of selected species in a variety of habitats. This would contribute to an understanding of the conditions under which species are able to survive in altered habitats. Better theoretical models of species viability are also needed in order to help scientists develop programs for saving and restoring endangered populations.[24] Needed as well is basic research in taxonomy, for otherwise how will we ever know how many species there really are?

Despite its shortcomings the Endangered Species Act is functioning reasonably well, and in practice its narrow legal focus on particular species (e.g., the spotted owl) often enlarges to a broader focus on related ecosystems (e.g., the redwood forest ecosystem of which the spotted owl is part). The act's success is a tribute to the pragmatism and shared social goals of our affluent society's various stakeholders—environmental groups, scientists, industrialists, government, and millions of citizens. Strong political opposition to the act remains, but its effect will probably be to improve the act rather than abolish it.

Looking to the developing countries, species preservation issues are often critically important because endangered species are so widely consumed in the private economies, especially by rural folk who actually use wild animal species in their daily lives. Although legislative equivalents to the Endangered Species Act are not likely in most developing countries, changes need to be brought about in the way species preservation issues are handled. Only when scientific approaches to biodiversity conservation are brought together with local culture and practice can conflicts be mitigated between preservationists, who advocate strict nonuse of threatened species, and conservationists, who advocate sustainable use of all species. Indeed, the top-down approaches to decision making that have been traditional among many conservation groups are now being supplemented by the active participation of local communities.

I noted earlier in this chapter that environmental pessimism tends to be particularly keen among biologists. This needs to be understood in context. Probably more than any other group, biologists embrace both the professional skills of science and an uncommon love and understanding of nature. Although aware that frequently they cannot provide solid scientific evidence to support their concerns about the environment, many biologists believe that they have an obligation to give timely warning of perceived environmental risks such as loss of biodiversity and its impact on future evolution. It was in that spirit that Rachel Carson wrote in *Silent Spring* about the possible harmful consequences of pesticide use, even though she went a bit far in her assertion that "for the first time in the history of the world, every human being is now subjected to contact with dangerous chemicals, from the moment of conception until death."[25] Actually, the vast bulk of the chemicals to which humans are exposed, and have always been exposed, are of natural origin.[26]

Despite their shortcomings, warnings such as *Silent Spring* helped to awaken a latent sensitivity to the environment in the affluent societies and also played a key role in the historical development of the environmental movement. People took these warnings seriously and reacted—sometimes overreacted—by supporting strong political and social actions to redress the environmental excesses wrought by industrialization. In the affluent countries the historical, and ongoing, record is full of examples of effective environmental legislation and regulation, which demonstrate the public's continuing commitment to the environment. The Endangered Species Act, a product of the world's most affluent society, is probably the clearest symbol of this commitment. That act and those that will follow in coming years demonstrate that the citizens of affluent and free societies are willing and able to take actions to preserve their habitat for future generations. Scientists have the difficult and often contentious task of clarifying the changing alternatives available to society, and these will always be clouded by the uncertainties inherent in empirical science.

But the story is different for people in the poorest places, such as the fishers who plunder the irreplaceable diversity of marine species in the Galápagos Islands. For them, *species diversity* most likely connotes being able to find enough plants and animals to provide food for next winter's table. As poverty diminishes, as people in the developing countries become more confident of a viable economic future for themselves and their descendants, they will increasingly appreciate their direct stake in the quality of the environment and the sustainable use of resources, and they will share those goals with the vast majority of the citizens of the affluent societies.

13

CHOICES

People everywhere care about their habitat. In the industrial countries, where freedom and affluence are the rule, people's social and political choices are generally friendly to the environment, and the goals of environmental quality enjoy broad public support. Because of that support, all the industrial countries have developed vigorous environmental programs, and many environmental success stories have been cited in these pages.

Success, however, needs to be judged in relation to expectations. As people become more affluent and more environmentally sensitive, their expectations of what constitutes a satisfactory environment constantly become loftier. By today's standards, yesterday's clean air would be considered polluted. *Thus environmental quality will always remain a work in progress.* One should not be surprised that the unprecedented affluence our society now enjoys brings not only environmental solutions but also new challenges—traffic congestion on city streets and highways, heavy use of public parks and open spaces, crowding of air lanes and airports—and policy dilemmas, especially the increasing demand for energy resources amidst increasing pressure to limit resource exploration. These problems can be vexing and divisive, but affluent democratic societies have the public support and political will, as well as institutions, to apply the best of science, technology, and management to find solutions.

Not so for the 80 percent of the world's people who have yet to attain affluence, and especially not for the unfortunate 20 percent still living in extreme poverty. For them, life's basic necessities, often survival itself, take on higher priority than environmental quality. Still, historical evidence supports the argument, made throughout these pages, that people and societies increasingly pursue environmental quality as they become less poor. This evidence is a major justification for optimism about the environmental future.

Optimism can help generate the human energy to surmount obstacles and the perseverance to find solutions. But optimism should never become a cover for complacency, for waiting passively until poverty disappears. The battle against poverty is far from won. Only an active commitment by individuals, institutions, and governments in the affluent world will ensure the global eradication of poverty. The moral case for this commitment is compelling. And from the perspective of the affluent countries, the pragmatic case is just as compelling, since failure to win the battle would almost certainly bring increasing environmental degradation, political insecurity, and disease throughout the world. Though it would be an exaggeration to link the September 11, 2001, terrorist attacks directly with poverty, a link certainly exists between the immense rich–poor gap and the festering disenchantment, humiliation, and hopelessness that together breed terrorism. Unfortunately the reality is that even in the best of circumstances, overcoming poverty may take several generations. During the long transition from poverty to affluence many of these ills will persist, including a significant legacy of environmental damage. We should strive to keep that damage to a minimum.

The true spirit of environmentalism embraces the twin goals of environmental improvement and poverty reduction. In the developing world, it is widely recognized that addressing the former without the latter would bring little reward. The influential 1987 Brundtland report on environment and development, for example, stressed the importance of economic growth in the developing countries as a prerequisite for environmental protection.[1] Yet some in the developing world still defend the traditional trappings of extreme poverty. Environmental activist Vandana Shiva, for example, describes subsistence farming in rural India as a cultural asset now under siege by the rich and powerful, and she attacks the remarkable global progress in food production with the claim that "the globalization of the food system is destroying the diversity of local food cultures and local food economies."[2] Those "local food cultures," unfortunately, were unable to feed India's poor children, one in three of whom died before age three, prior to the Green Revolution. If the glorification of traditional culture carries with it a glorification of poverty, this is no gift to the poor.

Whether in affluent or developing countries, the link between economic growth and environmental quality is vital. This link was recognized as a principle by the Brundtland report, and it has been confirmed by the actual experience of many affluent countries. In these pages the emphasis has been on the historical experience of the United States, whose robust economic growth and unequaled affluence have stimulated and supported ever stricter environmental protection, including measures such as the Clean Water Act, the Clean Air Act, vehicle fuel-efficiency standards, and the unique Endangered Species Act. Such environmental advances come out of affluence, not poverty.

Though the means to eradicate poverty are now attainable, actually doing so continues to be exceedingly challenging. But the battle is still young; only in the decade since the end of the Cold War has the global assault on poverty reached significant dimensions. During the Cold War the relationships between the great powers and the poor countries were, regrettably, determined more by geopolitical and strategic objectives than by goals of poverty reduction. Nonetheless some development-assistance efforts go back much further, at least to World War II. Most of the earlier efforts were focused on large infrastructure development projects, addressing general economic development rather than the specific needs of poor communities. International institutions such as the World Bank provided loans totaling well over a hundred billion dollars for such projects, including the construction of many large dams in developing countries. In all, about 540 large dams in ninety-two countries were built with World Bank loans.[3] Whether intended for electricity production, water supply, or irrigation, these dams made a positive contribution to development and economic progress. However, in many cases a large social and environmental price has been paid for the overall benefits of dams—dearly paid especially by the forty million to eighty million poor people who endured forced resettlement and loss of livelihood caused by the dams.[4] Although large infrastructure projects continue to be undertaken throughout the developing world—including the immense Three Gorges Dam in China, which may displace almost two million people—it is now widely recognized that many such projects were too narrowly focused and, while they often served the needs of the donors, they did not do enough to improve the lives of poor people.

In recent years, a sea change has taken place in the way aid and development are viewed. Both the causes and remedies of poverty are being addressed much more broadly by the global community. The following are increasingly recognized as fundamental goals for the elimination of poverty:

· Freedom and democracy are the sine qua non of the battle against poverty. They must become the universal right and achievement of all people and all nations.

· Gender equality must be achieved throughout the world. Poverty cannot be eradicated until women have equal opportunity to participate in their nations, their communities, their professions, and their families.

· Individual citizens of the developing world must be provided the tools to lift themselves from poverty, especially universal education of girls and boys, widely available public-health services, and quality medical care.

· New wealth must be created through economic growth that is both sustainable and equitable to the poor. The main engine of economic growth is enhanced human productivity, which can be achieved mainly through education, advances in scientific knowledge, and investment in and global diffusion of new technologies.

· A massive worldwide effort is needed to combat the diseases that cause or perpetuate poverty, especially malaria, tuberculosis, HIV/AIDS, and the childhood killer diseases. In Africa these diseases may be reducing economic growth by as much as half.

· The world's economy must become truly global, with all people and nations eligible and all enjoying a level playing field and fair international rules to protect the weak from the strong. Most important, the developing countries must have access to developed country markets.

· Foreign aid needs to be selectively targeted at the poorest countries and at those that have developed the most effective economic policies. A greater part of development assistance should be focused directly on the needs of the poor.

Indicative of the new thinking about poverty reduction, the World Bank and International Monetary Fund (IMF) recently launched a joint poverty-reduction initiative involving a much higher level of developing country participation than previous programs. The approach is based on the principle that each country should prepare and manage its own poverty-reduction plan, which presents the full set of options that the country intends to pursue to reduce poverty. Each country's plan is intended to serve as the basic framework for development assistance from the international donor community to that country. In the first three years of this program, about twenty countries have either completed or nearly completed their poverty-reduction strategy plans, while some forty other countries have prepared interim plans.[5]

These plans have shown some early success in many countries. Individual governments and their civil partners appear to be committed to the process. Issues related to poverty reduction are becoming more prominent in policy debates, and a more open dialogue is emerging within governments and with at least some parts of civil society. And many international donor organizations are actively helping countries prepare their strategic plans, in some cases through new or stronger partnerships.

But this new approach also brings new problems. For one thing, the process is extremely demanding, especially to countries with little tradition of assisting the poor and little experience in strategic planning. They must find consensus among a wide range of domestic stakeholders on complex issues including governance, macroeconomic policy, social inclusion, and public expenditures. They must decide which policies have the best chance of working, especially those aimed at promoting pro-poor economic growth and ensuring availability of quality services for the poor. They must become skilled in prioritizing their goals, in monitoring and evaluating results, and in making effective midcourse corrections. And not least, they must cope with the inevitable tensions between countries and donors. While donors obviously must remain accountable for the use of their resources, donor practices also need to empower governments to act on their own development strategies. In all these matters, coordination and flexibility are critical to success in these new partnerships between countries and donors.[6] As of now, it is much too early to predict how well the country plans will be implemented, or what their impacts will actually be on the poor.

The road to affluence is, lamentably, littered with the detritus of human history, culture, and oppression. Regardless of the quality of the new planning processes or the generosity of donors, the transition out of poverty will be a protracted and painful process for many developing countries. Individuals in leadership positions must defy tradition and develop the will and spirit to strive for the common good rather than private gain. Public service institutions must be established for which there is neither precedent nor experience. Ways must be found to overcome the stifling bureaucratic tendencies that inevitably develop in such institutions, eroding public trust and participation. Relations between representatives of donors and recipients will sometimes deteriorate disastrously because of cultural and linguistic barriers. Yet in spite of all these obstacles the battle against poverty is gaining momentum and the number of players keeps increasing.

For decades, a major player in the global war on poverty has been the United Nations. Recently the United Nations Development Programme (UNDP) has focused on the enormous impacts that technology can have

on development. In its 2001 report, the UNDP concludes that three fundamental developments of the new millennium will dramatically change the
ways technology can be used to help reduce poverty.

· INFORMATION AND COMMUNICATIONS. Almost everyone knows how
effectively information and communications technologies can increase
access to knowledge. These technologies are creating worldwide networks and portable devices that can bring everyone, rich and poor alike,
instantly in touch with everyone else, whether in nearby isolated villages or at the far corners of the earth. But they can also be of help
to poor people in several novel and specific ways. First, participatory
democracy has been given a tremendous boost by electronic mail, as
witness the recent (2001) avalanche of e-mail generated by citizens of
the Philippines during the impeachment trial of President Estrada. Second, governments and small businesses in developing countries are
increasingly able to use electronic databases to improve the efficiency
of their planning, budgeting, and management operations. Third, use of
the Internet helps increase profits by providing up-to-the-minute business information, for example, market prices and new techniques for
farmers and satellite-imaged shoal locations for fishers. Fourth, the ability of remote clinics to transmit medical data and digital images to diagnostic experts in distant medical centers is enabling quality treatment to
reach patients who previously had no access to modern medical care.

· BIOTECHNOLOGY. Chapter 3 reviewed the enormous potential of the
new genetically based biotechnologies to enable advances in agriculture and medicine. These could greatly enhance food security and health
care in the poorest countries, provided that the products of biotechnology become widely available in those countries. Critical contributions
may come from development of drought-tolerant and virus-resistant
varieties of the staple crops in sub-Saharan Africa and other marginal
lands. To this can be added the medical potential of biotechnology,
whose products have heretofore been available almost exclusively to
the rich countries. Possibilities exist for targeting the major health
challenges facing tropical countries with genetically engineered vaccines for malaria, HIV, tuberculosis, sleeping sickness, and river blindness. Genetic techniques may also make possible the introduction into
the tropics of mosquitoes that do not carry malaria. New diagnostic
methods and vaccines for preventing livestock diseases, including foot-
and-mouth disease, made possible through genetic engineering, would
be a boon to rich and poor countries alike.

· GLOBALIZATION. World trade has been a factor in the economic development of nations for centuries. Today, the term globalization refers to the rapid increases in the scale and importance of exchanges of people, products, services, capital, and ideas across international borders. The fundamental changes that have accelerated international exchanges are the falling costs of communications and transportation.[7] The economic globalization of world markets creates competition and incentives that greatly accelerate technological innovation in both the developed and developing worlds. Innovative technology increases productivity and economic growth, enabling people to lift themselves out of poverty.

Globalization is often the target for undeserved blame and credit. On one hand, it is blamed for the disturbing rise in wage inequality in the United States, but in fact the inequality arises mostly from advances in technology that disproportionately benefit more highly educated workers.[8] On the other hand, globalization is often given credit for the economic and political liberalization that is occurring around the world, but these trends reflect more the inexorable movement toward human freedom that has been occurring for more than a century.

In its 2001 development report, the UNDP established a set of ambitious "millennium" goals for worldwide poverty reduction, to be achieved by 2015.[9] A few examples of the UNDP goals and current progress toward them are as follows:

· GOAL: Reduce by half the proportion of people living in extreme poverty (defined as less than one dollar per day).

PROGRESS: Between 1990 and 1998 this proportion in developing countries was reduced from 29 percent to 24 percent, yet today 1.2 billion people still subsist in extreme poverty. China and India, with 38 percent of the world's people, are on track to meet the 2015 goal, as are nine other countries with 5 percent of the world's people. Yet seventy countries are far behind or slipping. Even if the goal is met, nine hundred million people will still be living in extreme poverty in 2015.

· GOAL: Reduce under-age-five mortality by two-thirds.

PROGRESS: Under-five mortality was reduced from ninety-three per one thousand live births to eighty in 1990–1999. Sixty-six countries are on track to meet the 2015 goal, yet ninety-three countries, with 62 percent of the world's population, are lagging the goal. Eleven million children under five still die each year from preventable causes.

· GOAL: Reduce by half the proportion of people without access to safe drinking water.

PROGRESS: Around 80 percent of people in the developing world now have access to improved water sources, yet nearly one billion still lack such access. Fifty countries are on track to reach the 2015 goal, but eighty-three countries with 70 percent of the world's population lag behind.

Achieving such goals comes with a heavy price tag. Some believe that double the current level of foreign aid will be required, and no clear plan has emerged for raising the $100 billion or so that may be needed annually. But beyond specific development goals and their financial implications, a new mood is arising among those who devote their lives to the battle against poverty. While recognizing the importance of economic growth in developing countries, the new viewpoint emphasizes that such growth should benefit the poor more directly. It places less emphasis on outsiders, especially government, doing things aimed at helping poor people and more emphasis on empowering poor people to do things to help themselves. Economic aid is a necessary part of such empowerment, of course, but the focus of action should be on creating a social climate enabling people to take more individual and collective responsibility for themselves. It should shift away from narrowly targeted government projects, toward improvements in national policy making and better governance of institutions devoted to poverty reduction. This viewpoint rejects the widespread belief that the poor are culturally less capable than others of providing for themselves, and it sets the norm of citizens themselves working collectively as prime movers in the fight against poverty, largely through self-organization at the community level.[10]

Debates about poverty often center on the roles of government and particularly on the question whether governments are typically more of a help or a hindrance to progress in poverty reduction. The positive roles of those governments with well-directed policies are real and widely acknowledged—for example, government-funded housing for the poor and government support for public health and women's education. But governments can also be an obstacle to poverty reduction. Political scientist Hernando de Soto points up the case of Peru, where the government allegedly denies poor people important property rights that the more affluent enjoy. Citing the example of Peruvian migrants who squat illegally on land they farm, Professor de Soto suggests that bureaucratic restrictions on land ownership, access to capital, and other legal rights severely hamper the economic productivity of which these and most other poor people are capable.[11] De Soto's view supports the increasingly held position that "poor people have talents and often know better what they need than outside consultants do."[12]

Solid and rapid economic growth probably remains the most effective social welfare policy, as indicated by the progress made against poverty in the fastest growing nations, especially China, whose income poor decreased from 260 million in 1978 to 42 million in 1998, despite the country's population growth.[13] In the developing countries as a whole, extreme poverty has been declining in recent decades, though the pace of decline has been slow and the trend erratic, with poverty actually rising in some periods even when regional economies were growing.[14] Still, most indicators confirm the overall decline of poverty in the developing countries. Since the 1960s, life expectancy has risen from forty-six to sixty-four years. Infant mortality rates have been cut in half. The proportion of children in primary schools has increased by 80 percent. Access to safe drinking water and basic sanitation has doubled.[15] The adult illiteracy rate has dropped from 31 percent in 1990 to 27 percent in 1997. The percentage of underweight children under five dropped from 33 percent in 1985 to 28 percent in 1995. Most other poverty indicators are also moving in the right direction.

Yet monetary wealth is only part of the story. As stressed by Amartya Sen (see Chapter 1), the economic aspect of poverty is but one facet of the underlying deprivation faced by poor people in many developing countries: their lack of basic human rights. The role of development should be not only to increase income but also to eliminate other deprivations, including tyranny, lack of health care and education, and denial of basic political and civil rights.[16] Eliminating these basic deprivations is the most fundamental aspect of the struggle against poverty. It is also a major ingredient of building a sustainable environment.

To individual citizens of the affluent countries, the battle against poverty often seems remote and beyond one's ability to influence. For those who have occasion to travel to the world's poorest places, the sense of remoteness is quelled by experiencing the environment of extreme poverty up close. Not so easily overcome is the sense of powerlessness to affect the problem. Yet there are things that individuals can do. Probably the most important is to support people's struggles everywhere for the basic freedoms. Through our choices at the ballot box, we can maintain pressure on our governments to invest more money in poverty reduction efforts, local and global. We can support nongovernmental organizations whose dedication to poverty reduction is explicit and evident, especially those that help people to help themselves. And we can support those leaders who have the optimism, vision, and will to persevere in finding solutions.

THE RICH, THE POOR, AND THE ENVIRONMENT

Over the past two centuries, industrial development paved the way out of poverty for millions. For most of those newly tasting affluence, the experience of an apparently unending stream of technological marvels was so bewitching that many traditional values became regarded as less important. Nature's bounty seemed infinitely abundant, with a cornucopia of natural resources and environmental services supporting the progress of technology. But nature found ways of demonstrating, mainly as a result of air and water pollution, that its environmental services must be protected if they are to be sustainable. To this day many of nature's environmental messages are incompletely understood by science or inaccurately transmitted by individuals and the media. Nonetheless, the affluent societies have responded with a spate of measures, some inconvenient and some costly, to protect their environment. And as affluence continues to grow, people's environmental expectations will inevitably also grow, as will the standards of environmental protection adopted in the affluent countries.

In contrast, the world's poor are struggling to escape from poverty, from tyranny, from ignorance, from dependence. These struggles are highly interconnected and increasingly bring conflict between poor and rich. Although environmental quality and protection of natural resources are not absent from the priority lists of the poor, they lie far below the other imperatives. If the affluent countries become earnest partners with the poor in their prime struggle against poverty, the poor countries will in turn become willing partners in the quest for a better global environment. 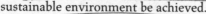 Through this partnership, and only through this partnership, can a truly sustainable environment be achieved.

NOTES

INTRODUCTION

1. A widely quoted definition of sustainable development is that given by the 1987 United Nations World Commission on Environment and Development, headed by Dr. Gro Harlem Brundtland: "growth that meets the needs of the present generation without compromising the ability of future generations to meet their needs."

2. The U.S. Environmental Protection Agency came into being on December 2, 1970, the first new line agency formed by the Nixon administration. The first administrator was William D. Ruckelshaus.

3. B. N. Ames and L. S. Gold, "Paracelsus to Parascience: The Environmental Cancer Distraction," *Mutation Research* 447 (2000): 3; idem, "Environmental Pollution, Pesticides, and the Prevention of Cancer: Misconceptions," *Federation of American Societies for Experimental Biology Journal* 11 (1997): 1041.

4. R. Carson, *Silent Spring* (Boston: Houghton Mifflin, 1962).

5. Committee on Research in the Life Sciences of the Committee on Science and Public Policy, *The Life Sciences: Recent Progress and Application to Human Affairs* (Washington, DC: U.S. National Academy of Sciences, 1970).

6. Before the ban was ordered by EPA, the judge who had conducted hearings on DDT concluded, "DDT is not a carcinogenic hazard to man. . . . The uses of DDT under the regulations involved here do not have a deleterious effect on freshwater fish, estuarine organisms, wild birds, or other wildlife. . . . The evidence in this proceeding supports the conclusion that there is a present need for the essential uses of DDT."

7. Environmental organizations including the World Wildlife Fund, Greenpeace, and Physicians for Social Responsibility urged the United Nations Environment Program (UNEP) to outlaw DDT worldwide. Many international organizations, including the Malaria Foundation, opposed the DDT ban, and an open letter protesting the ban was signed by 350 world experts in malaria,

including Nobelist Joshua Lederberg. At the UNEP's December 2000 meeting, 122 nations approved a Persistent Organic Pollutant Treaty that excluded DDT from the ban and provided for continued use of DDT "for public health purposes."

8. A. J. Lieberman and S. C. Kwon, *Facts versus Fear: A Review of the Greatest Unfounded Health Scares of Recent Times*, 3rd ed. (New York: American Council on Science and Health, 1998); N. Krieger et al., "Breast Cancer and Serum Organochlorines: A Prospective Study among White, Black, and Asian Women," *Journal of the National Cancer Institute* 86 (1994): 589.

9. M. L. Scott et al., "Effects of PCBs, DDT, and Mercury Compounds upon Egg Production, Hatchability and Shell Quality in Chickens and Japanese Quail," *Poultry Science* 54 (1975): 350; W. C. Krantz et al., "Organochlorine and Heavy Metal Residues in Bald Eagle Eggs," *Pesticides Monitoring Journal* 4(3) (1970): 136; W. Hazeltine, "Disagreements on Why Brown Pelican Eggs Are Thin," *Nature* 239 (1972): 410; E. S. Chang, and E. L. R. Stokstad, "Effect of Chlorinated Hydrocarbons on Shell Gland Carbonic Anhydrose and Egg-Shell Thickness in Japanese Quail," *Poultry Science* 54 (1975): 3. Also see the list of references compiled by J. G. Edwards and S. Milloy, *100 Things You Should Know about DDT*, at www.junkscience.com (1999).

10. F. L. Beebe, *The Myth of the Vanishing Peregrine* (N. Surrey, BC: Canadian Raptor Society Press, 1971); J. N. Rice, *Peregrine Falcon Populations* (Madison: University of Wisconsin Press, 1969), 155.

11. Advisory Committee on Toxic Chemicals, *Review of Organochlorine Pesticides in Britain* (Wilson Report) (Department of Education and Science, UK, 1969).

12. Garry Wills, *A Necessary Evil* (New York: Simon & Schuster, 1999).

13. This expression was taken from the title of the book by E. F. Schumacher, *Small Is Beautiful: Economics As If People Mattered* (London: Blond & Briggs, 1973).

14. In fact, the primary causes of the widespread gasoline shortages in the winter of 1973–1974 were a poorly functioning gasoline distribution system and consumer panic buying and hoarding.

15. Green Party Platform 2000, ratified at the Green Party National Convention, June 2000.

16. Albert Gore Jr., *Earth in the Balance: Ecology and the Human Spirit*, (Boston: Houghton Mifflin, 1992).

17. Gregg Easterbrook, "Green Surprise?" *Atlantic Monthly* (August 2000): 17.

18. To their credit, a few scientists have publicly confronted this issue; for example, this candid statement by climatologist Stephen Schneider: "On the one hand, as scientists we are ethically bound to the scientific method, in effect promising to tell the truth, the whole truth, and nothing but—which means that we must include all the doubts, the caveats, the ifs, ands and buts. On the other hand, we are not just scientists but human beings as well. And like most people we'd like to see the world a better place, which in this context translates

into our working to reduce the risk of potentially disastrous climate change. To do that we need to get some broad-based support, to capture the public's imagination. That, of course, entails getting loads of media coverage. So we have to offer up scary scenarios, make simplified, dramatic statements, and make little mention of any doubts we might have.... Each of us has to decide what the right balance is between being effective and being honest" (Schneider interview, *Discover* [October 1989]: 45).

19. This kind of process is sometimes described by economists in terms of an *Environmental Kuznets Curve*, and several economic studies have attempted to quantify the various stages of economic growth and environmental quality. See, for example, G. Grossman and A. Krueger, "Economic Growth and the Environment," *Quarterly Journal of Economics* 110(2) (1995): 353; T. M. Selden, and D. Song, "Environmental Quality and Development: Is There a Kuznets Curve for Air Pollutions Emissions," *Journal of Environmental Economics and Management* 27(2) (1994): 147.

20. N. E. Borlaug, *Feeding a World of 10 Billion People: The Miracle Ahead* (lecture given at De Montfort University, Leicester, UK, May 31, 1997).

21. World Bank, *Poverty Reduction and the World Bank: Progress in Fiscal 1996 and 1997* (Washington, DC: World Bank, 1998).

CHAPTER 1. A WORLD APART

1. The data in this section are taken from United Nations Food and Agricultural Organization, *The State of Food Insecurity in the World 1999* (Geneva: UNFAO, 1999).

2. Mark Hertsgaard, *Earth Odyssey* (New York: Broadway Books, 1998).

3. "Hopeless Africa," *Economist* (May 13, 2000): 17.

4. World Health Organization (WHO), *World Health Report, 2000* (Geneva: WHO, June 4, 2000).

5. Ibid.

6. Data from United Nations International Children's Emergency Fund, *The State of the World's Children 1996* (New York: UNICEF, 1996).

7. See, for example, South Coast Air-Quality Management District, *1999 Current Air-Quality and Trends* (Diamond Bar, CA, May 2000).

8. Amartya Sen, *Development as Freedom* (New York: Knopf, 1999).

CHAPTER 2. SIX BILLION AND COUNTING

1. J. E. Cohen, *How Many People Can the Earth Support?* (New York: W. W. Norton, 1995).

2. William Petty, *The Economic Writings of Sir William Petty*, ed. C. H. Hull (1676; reprint, Cambridge, UK: Cambridge University Press, 1899).

3. Julian L. Simon, *The Ultimate Resource 2* (Princeton, NJ: Princeton University Press, 1996).

4. Soichiro Honda, *Wall Street Journal*, February 1, 1982, 15, quoted in Simon, *Ultimate Resource 2*, 380.

5. Christopher Cerf and Victor Navasky, *The Experts Speak* (New York: Pantheon Press, 1984).

6. T. Malthus, *An Essay on the Principles of Population as It Affects the Future Improvement of Society* (London: J. Johnson, 1798).

7. G. Ortes, quoted in *Economist* (December 31, 1999).

8. P. R. Ehrlich, *The Population Bomb* (New York: Ballantine Books, 1968); P. R. Ehrlich and A. Ehrlich, *Population Explosion* (New York: Simon & Schuster, 1990).

9. United Nations, *World Population Prospects: The 1998 Revision* (New York: UN Population Division, Dept. of Economic and Social Affairs, 1998).

10. United Nations, *World Population Prospects: The 2000 Revision* (New York: UN Population Division, Dept. of Economic and Social Affairs, 2000).

11. United Nations, *World Population Prospects: The 1998 Revision*.

12. U.S. Department of Commerce, *U.S. Statistical Tables* (Washington, DC, 1996).

13. When a country's fertility rate drops to or below the replacement level, a time lag occurs before the population actually declines, because of the large numbers of women who are in (or approaching) their child-bearing years. Demographers call this phenomenon "population momentum."

14. These and related factors are reviewed by Vernon W. Ruttan in "Perspectives on Population and Development," *Indian Journal of Agricultural Economics* 39(4) (October–December 1984): 630–638.

15. Biologist Garrett Hardin has been a leading proponent of coercive policies for population control. See his article "Tragedy of the Commons," *Science* 162 (1968): 1243. The organization Zero Population Growth has long advocated voluntary family planning programs.

16. C. A. Scotese and P. Wang, "Can Government Enforcement Permanently Alter Fertility? The Case of China," *Economic Inquiry* (October, 1995): 552.

17. United Nations, *World Population Prospects: The 1998 Revision*.

18. Statistics Sweden, *Sweden's Future Population* (Stockholm: Statistiska Centralbyran, 1994).

19. United Nations, *World Population Prospects: The 2000 Revision*.

CHAPTER 3. CAN THE EARTH FEED EVERYONE?

1. N. Sharma, ed., *Managing the World's Forests* (Washington, DC: World Bank, 1992).

2. S. Vosti, *The Role of Agriculture in Saving the Rain Forest*, 2020 Vision Brief 9 (Washington, DC: International Food Policy Research Institute, February 1995).

3. Note a difference between this and the energy resource problem. Increasing uses of coal and oil by developing countries typically create pollution problems that are mostly reversible, as discussed in Chapter 7, but are not

likely to create long-term supply constraints, since the availability of energy resources is expected to be adequate for the foreseeable future.

4. T. Malthus, *An Essay on the Principles of Population as It Affects the Future Improvement of Society* (London: J. Johnson, 1798).

5. P. R. Ehrlich, *The Population Bomb* (New York: Ballantine Books, 1968); P. R. Ehrlich, and A. Ehrlich, *Population Explosion* (New York: Simon & Schuster, 1990).

6. C. C. Mann, "Crop Scientists Seek a New Revolution," *Science* 283 (January 15, 1999): 310.

7. Norman E. Borlaug was awarded the Nobel Peace Prize in 1970 to honor his leading contribution to the Green Revolution and to acknowledge the millions of lives that have been saved by this work.

8. R. Bailey, "Billions Served" (interview with Norman Borlaug), *Reason* (April 2000), www.reason.com; N. E., Borlaug, *Feeding a World of 10 Billion People: The Miracle Ahead* (lecture given at de Montfort University, Leicester, UK, May 31, 1997).

9. Mann, "Crop Scientists Seek a New Revolution."

10. Among the organizations most active in this work are the Consultative Group on International Agricultural Research and the International Food Policy Research Institute, both based in Washington, DC

11. M. J. Cohen and D. Reeves, *Causes of Hunger,* 2020 Vision Brief 19 (Silver Spring, MD: International Food Policy Research Institute, May 1995).

12. The term *calories* is used here in deference to common practice, as for example, on food labels. It is, however, a misnomer; the correct, scientific unit is the kilocalorie, that is, a thousand calories—sometimes called a "large calorie."

13. Food and Agricultural Organization, *Agriculture: Towards 2015/30* (Geneva: United Nations Food and Agricultural Organization, 2000).

14. United Nations, *World Population Prospects: The 2000 Revision* (New York: UN Population Division, Dept of Economic and Social Affairs, 2000).

15. P. E. Waggoner, "How Much Land Can Be Spared for Nature?" *Daedalus* (special issue, ed. J. Ausubel) 125(3) (summer 1996).

16. K. G. Cassman, "Ecological Intensification of Cereal Production Systems: Yield Potential, Soil Quality, and Precision Agriculture," *Proceedings of the National Academy of Sciences* 96 (May 1999): 5952.

17. J., Macedo, *Prospects for the Rational Use of the Brazilian Cerrado for Food Production* (Brazilian Agricultural Research Corporation, CPAC/EMBRAPA, Brasilia DF, 1995), cited in Borlaug, *The Miracle Ahead.*

18. Waggoner, "How Much Land Can Be Spared."

19. C. Delgado, M. Rosegrant, H. Steinfeld, S. Ehui, and C. Courbois, *Livestock to 2020: The Next Food Revolution,* 2020 Vision Brief 61 (Washington, DC: International Food Policy Research Institute, 1999).

20. W. Bender, *An End Use Analysis of Global Food Requirements,* Food Policy Statement 19, no. 4 (Silver Spring, MD: International Food Policy Research Institute, 1994).

21. Ibid.

22. Borlaug, *The Miracle Ahead*.

23. P. Pinstrup-Andersen, R. Pandya-Lorch, and M. W. Rosegrant, *World Food Prospects: Critical Issues for the Early Twenty-First Century*, 2020 Vision Food Policy Report (Washington, DC: International Food Policy Research Institute, October 1999).

24. I am indebted to Drs. Marc J. Cohen and Don Reeves, from whose analysis of this issue I have liberally drawn. See Cohen and Reeves, *Causes of Hunger*, 2020 Vision Brief 19; idem, *Causes of Hunger*, 2020 Vision Brief 29 (Washington, DC: International Food Policy Research Institute, May 1995).

25. World Bank, *1997 World Development Indicators* (Washington, DC: World Bank, 1997).

26. A. F. McCalla, *The Challenge of Food Security in the 21st Century* (address given at McGill University, Montreal, June 5, 1998).

27. Cohen and Reeves, *Causes of Hunger*, 2020 Vision Brief 29.

28. Ibid.

29. J. Von Braun, T. Teklu, and P. Webb, *Famine in Africa* (Baltimore: Johns Hopkins University Press, 1999).

30. Dividing people into "optimists" and "pessimists" is obviously an oversimplification of the breadth of human motivations and value judgments, but it does indicate how very different conclusions can be reached, even by scientists, from identical information.

31. G. Daily et al. (fifteen other authors), "Food Production, Population Growth, and the Environment," *Science* 281 (1998): 1291.

32. Mann, "Crop Scientists Seek a New Revolution."

33. International Food Policy Research Institute, *The World Food Situation: Recent Developments, Emerging Issues, and Long-Term Prospects* (Washington, DC: IFPRI, 1997).

34. R. L. Naylor, "Energy and Resource Constraints on Intensive Agricultural Production," in *Annual Review of Energy and the Environment*, vol. 21, ed. R. Socolow (Palo Alto, CA: Annual Reviews, 1996), 99.

35. Examples are human insulin; dornase alpha, a breakthrough treatment for cystic fibrosis; interferon beta, a powerful drug for certain multiple sclerosis cases; activase, a clot-dissolving agent used to treat heart attacks; and a synthetic hepatitis B vaccine free of human blood infections.

36. Borlaug, *The Miracle Ahead*.

37. A. Sittenfeld, A. M. Espinoza, M. Munoz, and A. Zamora, "Costa Rica: Challenges and Opportunities in Biotechnology and Biodiversity," in *Agricultural Biotechnology and the Poor*, ed. G. J. Persley and M. M. Lantin (Washington, DC: Consultative Group on International Agricultural Research, 2000), 79.

38. G. J. Persley, "Agricultural Biotechnology and the Poor: Promethean Science," in *Agricultural Biotechnology and the Poor*, 3.

39. D. Normile, "Crossing Rice Strains to Keep Asia's Rice Bowls Brimming," *Science* 283 (January 15, 1999): 313.

40. Sittenfeld et al., "Costa Rica."

41. M. Tanticharoen, "Thailand: Biotechnology for Farm Products and Agro-Industries," in *Agricultural Biotechnology and the Poor,* 64.

42. D. Gonsalves, "Control of Papaya Ringspot Virus in Papaya: A Case Study," *Annual Review of Phytopathology* 36 (1998): 415.

43. M. J. A. Sampaio, "Brazil: Biotechnology and Agriculture to Meet the Challenges of Increased Food Production," in *Agricultural Biotechnology and the Poor,* 74.

44. C. James and A. Krattiger, "The Role of the Private Sector," brief 4 in *Biotechnology for Developing-country Agriculture: Problems and Opportunities,* Focus 2: 2020 Vision, ed. G. Persley (Washington, DC: International Food Policy Research Institute, October 1999).

45. A. McHughen, *Biotechnology and Food* (New York: American Council on Science and Health, September 2000).

46. Working Group of Academies of Sciences, *Transgenic Plants and World Agriculture* (Washington, DC, National Academy Press, July 2000); McHughen, *Biotechnology and Food.*

47. McHughen, *Biotechnology and Food.*

48. Turning Point Project, *Genetic Roulette* (advertisement no. 3 in a series on genetic engineering) (Washington, DC: TPP, 2000).

49. McHughen, *Biotechnology and Food.*

50. Associated Press (AP), "Monarch Butterflies Abundant in Mexican Sanctuaries," Nando Media Online (November 1, 1999). See also T. R. DeGregori, *Genetically Modified Nonsense* (London: Institute of Economic Affairs, January 1, 2001), www.iea.org.uk.

51. C. S. Prakash et al., "Declaration of Scientists in Support of Agricultural Biotechnology," www.agbioworld.org/petition.phtml (2000).

52. N. Smith, *Seeds of Opportunity: An Assessment of the Benefits, Safety and Oversight of Plant Genomics and Agricultural Biotechnology,* report prepared for the Subcommittee on Basic Science of the House Committee on Science, 106th Cong., 2d sess., 2000, www.house.gov/science.

53. National Research Council, *Report of Committee on Genetically Modified Pest-Protected Plants* (Washington, DC: National Academy of Sciences, 2000).

54. An example is this header from a newspaper advertisement placed by the anti-technology lobby group, Turning Point: "The genetic structures of living things are the last of Nature's creations to be invaded and altered for commerce. Now they're being seized for corporate ownership. Nothing will ever be the same, and we approach the gravest moral, social, and ecological crisis in history."

55. The anti-technologists would probably respond that taking the risk of automobile travel is voluntary, whereas eating genetically engineered food involves an involuntary risk. However, in the industrialized world, there are few substitutes for the automobile in contemporary life, so the distinction here becomes essentially moot.

56. Borlaug, *The Miracle Ahead.*

CHAPTER 4. FISH TALES

1. G. Hardin, "The Tragedy of the Commons," *Science* 162 (1968): 1243.

2. In the mid-1970s, Exclusive Economic Zones were established, under the UN Convention on the Law of the Sea, that give most coastal nations exclusive fishing rights within two hundred miles from their shores.

3. UN Food and Agricultural Organization, *Review of the State of World Marine Fishery Resources,* Technical Paper 335 (Rome: UNFAO, 1994).

4. L. W. Botsford, J. C. Castilla, and C. H Peterson, "The Management of Fisheries and Marine Ecosystems," *Science* 277 (July 25, 1997): 509.

5. UN Food and Agricultural Organization, "World Fisheries," chapter 7 in *Agriculture: Towards 2015/30* (Rome: UNFAO, 2000). See also UNFAO, *FAO Yearbook: Fishery Statistics—Catches and Landings 1993*, Vol. 76 (Rome: UNFAO, 1995).

6. H. Shand, *Human Nature: Agricultural Biodiversity and Farm-Based Food Security* (Pittsboro, NC: Rural Advancement Foundation International, 1997).

7. B. J. Rothschild, *How Bountiful Are Ocean Fisheries? Consequences* 2(1) (1996); 15; P. Weber, *Net Loss: Fish, Jobs, and the Marine Environment*, Worldwatch paper 120 (Washington, DC: Worldwatch Institute, 1994): 16.

8. E. Ostrom, *Governing the Commons: The Evolution of Institutions for Collective Action* (Cambridge, UK: Cambridge University Press, 1990).

9. R. S. Steneck, "Human Influences on Coastal Ecosystems: Does Over-Fishing Create Trophic Cascades?" *Tree* 13(11) (1998): 429.

10. E. Sala, C. F. Boudouresque, and M. Harmelin-Vivien, "Fishing, Trophic Cascades, and the Structure of Algal Assemblages: Evaluation of an Old but Untested Paradigm," *Oikos* 82 (1998): 425.

11. Botsford, Castilla, and Peterson, "Management of Fisheries."

12. E. L. Venrick, J. A. McGowan, D. R. Cayan, and T. L. Hayward, "Climate and Chlorophyll a: Long-Term Trends in the Central North Pacific Ocean," *Science* 238 (1987): 70.

13. W. G. Pearcy, *Ocean Ecology of North Pacific Salmonids* (Seattle: University of Washington Press, 1992).

14. D. L. Alverson, *Review of Aquatic Science* 6 (1992): 203.

15. D. Lluch-Beldaet et al., *South African Journal of Maritime Science* 8 (1989): 195; T. Kawasaki et al., eds., *Long-Term Variability of Pelagic Fish Populations and Their Environment* (New York: Pergamon Press, 1991).

16. Botsford, Castilla, and Peterson, "Management of Fisheries."

17. *Aquiculture* refers to all types of fish farming in enclosures such as ponds, tanks, and pens, as well as culture-based fisheries. Fish farming implies individual or corporate ownership of the stock being cultivated, as opposed to fisheries, in which aquatic organisms are exploitable by the public as a common property resource, with or without appropriate licenses.

18. M. De Alessi, "Fishing for Solutions," in *Earth Report 2000*, ed. R. Bailey (New York: McGraw Hill, 2000).

19. UN Food and Agricultural Organization, *Review of the State of World Marine Fishery Resources.*

20. P. Edwards, "Aquaculture, Poverty Impacts, and Livelihoods," in *Natural Resources Perspectives,* ed. J. Farrington (London: Overseas Development Institute, 2000).

21. P. Edwards, *A Systems Approach for the Promotion of Integrated Aquaculture* (paper presented at Integrated Fish Farming International Workshop, Wuxi, People's Republic of China, October 1994).

22. Edwards, "Aquaculture, Poverty Impacts, and Livelihoods."

23. S. Jentoft and T. Kristoffersen, "Fishermen's Co-Management: The Case of the Lofoten Fishery," *Human Organization* 48(4) (1989): 355.

24. D. R. Leal, "Community-Run Fisheries: Avoiding the 'Tragedy of the Commons,'" *Political Economy Research Series,* Issue PS-7, ed. J. S. Shaw (Bozeman, MT: Political Economy Research Center, 1996).

25. Ostrom, *Governing the Commons.*

26. B. Runolfsson, "Fencing the Oceans: A Rights-Based Approach to Privatizing Fisheries," *Regulation* 20(3) (summer 1997): 57.

27. Ibid.

28. Ibid.

29. S. H. Verhovek, "Returning River to Salmon, and Man to the Drawing Board," *New York Times,* September 26, 1999, 1-1.

30. S. Gorton, quoted in Verhovek, "Returning River to Salmon."

31. B. Finney et al., "Impacts of Climate Change and Fishing on Pacific Salmon Abundance over the Past 300 Years," *Science* 290 (October 27, 2000), 795.

CHAPTER 5. IS THE EARTH WARMING?

1. More precisely, the earth's *surface* warmed about 0.4 degree Celsius between 1860 and 1940, cooled about 0.1 degree Celsius from 1940 to 1980, and warmed about 0.3 degree in the past two decades. But temperature measurements taken from satellites do not show any warming of the earth's *atmosphere* in the past two decades. See this chapter for details.

2. P. N. Edwards, and S. H. Schneider, "Self-Governance and Peer Review in Science-for-Policy: The Case of the IPCC Second Assessment Report, in *Changing the Atmosphere; Expert Knowledge and Environmental Governance,* ed. C. Miller and P. N. Edwards (Cambridge, MA: MIT Press, 2001). This is an unnatural hybrid. True scientific organizations function by furthering the dissemination of original research, which becomes "knowledge" after passing through the fine filter of replication, discussion, and debate and earns eventual acceptance or rejection by the scientific community. Scientific organizations do not determine truth by processes of negotiation and vote, as do political organizations. The processes by which IPCC has determined its policy positions establish it as more a political than a scientific organization. See John Ziman, *Public Knowledge: An Essay Concerning the Social Dimension of Science* (Cambridge, UK: Cambridge University Press, 1968).

3. C. A. Miller and P. N. Edwards, introduction to *Changing the Atmosphere.*

4. B. McKibben, "A Special Moment in History," *Atlantic Monthly* (May 1998): 55.

5. Conference of the Parties 3 (COP-3), *Kyoto Protocol to the United Nations Framework Convention on Climate Change* (Kyoto, Japan: December 1997).

6. Sierra Club, Global Warming (March 1999), www.sierraclub.org/global warming.

7. In accepted usage the term *pollutant* refers to substances introduced into the environment that are known to have negative consequences for health or welfare. Carbon dioxide does not have such negative consequences; hence it is not subject to regulation by the Environmental Protection Agency as a pollutant. In fact carbon dioxide is a beneficial substance, vital to all life.

8. Greenhouse gases acting alone, however, would make the earth too hot. There are also natural cooling processes, such as evaporation of water, that balance the heating from the greenhouse gases and produce the favorable earth temperatures that we enjoy.

9. Frank Shu, *The Physical Universe: An Introduction to Astronomy* (Herndon, VA: University Science Books, 1982).

10. The quoted value is from measurements at Mauna Loa, Hawaii. Other measuring sites give somewhat different values at any given time, but the overall trends are consistent.

11. World Resources Institute, *World Resources, 2000–2001* (Oxford: Oxford University Press, 2000), data table AC.3.

12. Svante Arrhenius, "On the Influence of Carbonic Acid in the Air upon the Temperature of the Ground," *Philosophical Magazine* 41 (1896): 237.

13. One of the most remarkable methods that scientists use for studying the earth's climate history is the drilling of ice cores thousands of feet into the earth's ice sheets. Like tree rings, each annual layer of ice provides direct evidence of volcanic eruptions, pollution, dust storms, and indirect evidence of air temperature through observed ratios of oxygen isotopes in the frozen water molecules. Ice-core measurements give data from periods as far back as millions of years—long before humans and thermometers existed.

14. J. Hansen, R. Ruedy, J. Glascoe, and M. Sato, "GISS Analysis of Surface Temperature Change," *Journal of Geophysical Research* 104 (1999): 30997.

15. R. Balling Jr., "Global Warming: Messy Models, Decent Data, and Pointless Policy," in *The True State of the Planet,* ed. R. Bailey (New York: Free Press, 1995), 83–107.

16. A. Henderson-Sellers and P. J. Robinson, *Contemporary Climatology* (New York: John Wiley, 1986). This cooling trend can be partially explained by the large increase in sulfate aerosol (tiny particle) emissions from coal burning over that period. But sulfate aerosols are only one of several kinds of aerosols that contribute to heating and cooling the atmosphere, in ways that are not presently well understood. In addition, other natural causes of cooling cannot be ruled out. See V. Ramanathan et al., "Aerosols, Climate, and the Hydrologi-

cal Cycle," *Science* 294 (December 7, 2001): 2119. Also see discussion of aerosols later in this chapter.

17. National Academy of Sciences–National Research Council Committee, *Understanding Climate Change: A Program for Action* (Washington, DC: NAS-NRC, 1975); S. H. Schneider, *The Genesis Strategy: Climate and Global Survival* (New York: Plenum Press, 1976).

18. Freeman Dyson, personal communication (May 14, 2001).

19. National Research Council, *Reconciling Observations of Global Temperature Change* (Washington, DC: National Academy Press, 2000).

20. Ibid.

21. J. Hansen, R. Ruedy, J. Glascoe, and M. Sato, "GISS Analysis of Surface Temperature Change."

22. J. Hansen et al., "A Closer Look at United States and Global Surface Temperature Change," *Journal of Geophysical Research,* 106(D20) (October 2001): 23947.

23. R. A. Kerr, "From Eastern Quakes to a Warming's Icy Clues," *Science* 283 (January 1, 1999): 29.

24. T. R. Naish, et al. (thirty-two coauthors), *Nature* 413 (October 18, 2001): 719.

25. C. A. Perry, and K. J. Hsu, *Proceedings of the National Academy of Sciences* (early edition) (September 2000).

26. Ray Bradley, "1000 Years of Climate Change," *Science* 288 (May 26, 2000): 1353; L. D. Keigwin, "The Little Ice Age and Medieval Warm Period in the Sargasso Sea," *Science* 274 (November 29, 1996): 1504.

27. Bradley, "1000 Years of Climate Change."

28. K. R. Briffa, P. D. Jones, F. H. Schweingruber, and T. J. Osborn, "Influence of Volcanic Eruptions on Northern Hemisphere Summer Temperature over the Past 600 Years," *Nature* 393 (1998): 450; P. D. Jones, K. R. Briffa, T. P. Barnett, and S. F. B. Tett, "High-Resolution Paleoclimatic Records for the Last Millennium: Interpretation, Integration, and Comparison with General Circulation Model Control-Run Temperatures," *Holocene* 8 (1998): 455; M. E. Mann, R. S. Bradley, and M. K. Hughes, "Global-Scale Temperature Patterns and Climate Forcings over the Past Six Centuries," *Nature* 392 (1998): 779; J. Overpeck et al., "Arctic Environmental Changes of the Last Four Centuries," *Science* 278 (1997): 1251.

29. Various authors in seven separate papers, "PaleoClimate," *Science* 292 (April 27, 2001): 657.

30. J. Smith and J. Uppenbrink, introduction to "PaleoClimate," 657.

31. United Nations Intergovernmental Panel on Climate Change, *Climate Change 2001: Summary for Policymakers* (Cambridge, UK: Cambridge University Press, 2001).

32. National Research Council, *Global Temperature Change.*

33. Ramanathan et al., "Aerosols, Climate, and the Hydrological Cycle," 2119.

34. UN Intergovernmental Panel on Climate Change, *Technical Summary of the Working Group I Report* (Geneva: IPCC, 2001).

35. J. Hansen, M. Sato, R. Ruedy, A. Lacis, and V. Oinas, "Global Warming in the 21st Century: An Alternative Scenario," *Proceedings of the National Academy of Sciences* 97(18) (2000): 9875.

36. Ramanathan et al., "Aerosols, Climate, and the Hydrological Cycle."

37. Sallie Baliunas, testimony before Senate Committee on Environment and Public Works, 106th Cong., 2d sess., March 13, 2002.

38. B. D. Santer et al., "Uncertainties in Observationally Based Estimates of Temperature Change in the Free Atmosphere," *Journal of Geophysical Research*, 104 (March 27, 1999): 6305.

39. J. Hansen, "Climate Forcings in the Industrial Era," *Proceedings of the National Academy of Sciences* 95(22) (October 27, 1998).

40. D. Rind, "Complexity and Climate," *Science* 284 (April 2, 1999): 105.

41. UN Intergovernmental Panel on Climate Change, *Climate Change 1995: Impacts, Adaptations and Mitigation: Summary for Policymakers* (Cambridge, UK: Cambridge University Press, 1996).

42. Committee on the Science of Climate Change, NAS-NRC, *Climate Change Science. An Analysis of Some Key Questions* (Washington, DC: National Academy Press, 2001).

43. M. Parry, N. Arnell, M. Hulme, R. Nicholls, and M. Livermore, "Adapting to the Inevitable," *Nature* 395 (1998): 741; D. Malakoff, "Thirty Kyotos Needed to Control Warming," *Science* 278 (1997): 2048.

44. Baliunas, testimony before Senate Committee on Environment and Public Works.

45. *Kyoto Protocol to the United Nations Framework Convention on Climate Change* (Bonn, Germany: UNFCCC, December 1997).

46. The Senate resolution, sponsored by Senators Chuck Hegel of Nebraska and Robert Byrd of West Virginia, passed in July 1997 by a unanimous vote of 95–0.

47. George W. Bush, President's address on global climate change, White House News Release, Office of Press Secretary, June 11, 2001.

48. W. D. Nordhaus, "Global Warming Economics," *Science* 294 (November 9, 2001): 1283.

49. L. D. Keigwin, "The Little Ice Age and Medieval Warm Period in the Sargasso Sea."

50. W. D. Nordhaus, "To Slow or Not to Slow: The Economics of the Greenhouse Effect," *Economics Journal* 101 (July 1991): 920.

51. S. H. Wittwer, *Food, Climate, and Carbon Dioxide* (Boca Raton, FL: CRC Press, 1995).

52. R. Mendelsohn and J. E. Neumann, eds., *The Impact of Climate Change on the United States Economy* (Cambridge, UK: Cambridge University Press, 1999).

53. UN Intergovernmental Panel on Climate Change (IPCC), *Summary for Policymakers* (1995), 13.

54. G. Taubes, "Apocalypse Not," *Science* 278 (1997): 1004.

55. Paul Reiter, *Global Warming and Vector-Borne Disease: Is Warmer Sicker?* (briefing for the National Consumer Coalition), http://www.cei.org (July 28, 1998).

56. Meteorological Office, Department of the Environment, United Kingdom, *Climate Change and Its Impacts* (London: November 1998).

57. UN IPCC, "Changes in Sea Level," chapter 11 in *Climate Change 2000: Summary for Policymakers* (Cambridge, UK: Cambridge University Press, 2000).

58. UN IPCC, "Changes in Sea Level."

59. J. L. Daly, *Testing the Waters: A Report on Sea Levels* (Arlington, VA: Greening Earth Society, 2000).

60. J. A. Dowdeswell et al., "The Mass Balance of Circum-Arctic Glaciers and Recent Climate Change," *Quaternary Research* 48 (1997): 1.

61. H. Conway, B. L. Hall, G. H. Denton, A. M. Gades, and E. D. Waddington, "Past and Future Grounding-Line Retreat of the West Antarctic Ice Sheet," *Science* 286 (1999): 280.

62. A. Trupin and J. Wahr, "Spectroscopic Analysis of Global Tide Gauge Sea-Level Data," *Geophysical Journal International* 100 (1990): 441.

63. These studies are summarized in D. Ridenour, *Don't Like the Weather? Don't Blame It on Global Warming*, Policy Analysis no. 206 (Washington, DC: National Center for Public Policy Research, August 1998); G. Van der Vink et al., "Why the United States Is Becoming More Vulnerable to Natural Disasters," *EOS: Transactions of the American Geophysical Union* 79(44) (November 3, 1998): 533.

64. Van der Vink, "Why the United States Is Becoming More Vulnerable to Natural Disasters."

65. W. D. Nordhaus, "An Optimal Transition Path for Controlling Greenhouse Gases," *Science* 258 (November 20, 1992): 1315; S. Fankhauser, "The Economic Costs of Global Warming: A Survey," *Global Environmental Change* 4 (December 1994): 301–309.

66. Mendelsohn and Neumann, *Impact of Climate Change.*

67. Nordhaus, "Global Warming Economics."

68. Five-Laboratory Working Group, *Scenarios of U.S. Carbon Reductions* (Washington, DC: U.S. Department of Energy, 1997).

69. J. Kaiser, "Pollution Permits for Greenhouse Gases?" *Science* 282 (1998): 1024.

70. A. Z. Rose and G. Oladosu, "Greenhouse Gas Reduction Policy in the United States: Identifying Winners and Losers in an Expanded Permit Trading System," *Energy Journal* 23(1) (January 2002): 1–18.

71. W. Beckerman, *Through Green-Colored Glasses* (Washington, DC: Cato Institute, 1996), 113.

72. H. Jacoby, R. Prinn, and R. Schmalensee, "Kyoto's Unfinished Business," *Foreign Affairs* (July–August 1998): 54–66.

CHAPTER 6. WATER, WATER EVERYWHERE

1. Paul Simon, *Tapped Out* (New York: Welcome Rain Publishers, 1998).

2. The water table is the level to which water will rise in an open well.

3. Peter H. Gleick, *The World's Water, 2000–2001* (Washington, DC: Island Press, 2000), 22.

4. S. L. Postel, G. C. Daily, and P. R. Ehrlich, "Human Appropriation of Renewable Fresh Water," *Science* 271 (February 9, 1996): 785.

5. I. A. Shiklomanov, *Assessment of Water Resources and Water Availability in the World*, report for the Comprehensive Global Freshwater Assessment of the United Nations (St. Petersburg, Russia: State Hydrological Institute, 1996), quoted by Gleick, *The World's Water* (Washington, DC: Island Press, 1998).

6. Gleick, *The World's Water, 2000–2001*, table 3.15.

7. United Nations, *World Population Prospects: The 1998 Revision* (New York: UN Population Division, Dept. of Economic and Social Affairs, 2000).

8. R. A. Downing, *Groundwater: Our Hidden Asset*, Earthwise Series (Keyworth, Nottingham, UK: British Geological Survey, 1998).

9. World Resources Institute, *World Resources, 1998–1999* (Washington DC, 1998), 304, table 12.1.

10. M. W. Rosengrant, *Dealing with Water Scarcity in the Next Century*, 2020 Vision Brief 21 (Washington, DC: International Food Policy Research Institute, 1995).

11. United Nations, *Comprehensive Assessment of the Freshwater Resources of the World* (New York: UN Commission on Sustainable Development, February 1997).

12. Gleick, *The World's Water, 2000–2001*, fig. 5.3 and table 20.

13. T. L. Anderson, "Water Options for the Blue Planet," in *The True State of the Planet*, ed. Ronald Bailey (New York: Free Press, 1995), chapter 8.

14. Centre for Natural Resources, United Nations, *Registry of International Rivers* (New York: Pergamon Press, 1978).

15. M. Falkenmark, "Fresh Waters as a Factor in Strategic Policy and Action," in *Global Resources and International Conflict: Environmental Factors in Strategic Policy and Action*, ed., A. H. Westing (New York: Oxford University Press, 1986), 85, cited by Gleick, *The World's Water* (1998), 108.

16. International Conference on Water and the Environment, Dublin (June 1992).

17. M. Xie, U. Kuffner, and G. Le Moigne, *Using Water Efficiently*, World Bank Technical Paper 205 (Washington, DC: World Bank, 1993).

18. S. Postel, *Last Oasis: Facing Water Scarcity* (New York: W. W. Norton, 1992, 1997).

19. Gleick, *The World's Water, 2000–2001*, 80.

20. M. W. Rosengrant, *Dealing with Water Scarcity in the Next Century*, 2020 Vision Brief 21 (Washington, DC: International Food Policy Research Institute, 2000).

21. R. L. Snyder, M. A. Plas, and J. I. Grieshop, "Irrigation Methods Used in California: Grower Survey," *Journal of Irrigation and Drainage Engineering* 122 (July–August 1996): 259; Peter H. Gleick, "Crop Shifting in California: Increasing Farmer Revenue, Decreasing Farm Water Use," in *Sustainable Use of Water: California Success Stories,* ed. L. Owens-Viani, A. K. Wong, and P. H. Gleik (Oakland: Pacific Institute, 1999), 149.

22. S. Postel, "Increasing Water Efficiency," in *State of the World, 1986,* ed. Worldwatch Institute (New York: W. W. Norton, 1986).

23. Gleick, *The World's Water* (1998), 24.

24. M. Reisner, *Cadillac Desert: The American West and Its Disappearing Water* (New York: Penguin Books, 1986; rev. ed., 1993).

25. World Commission on Water for the 21st Century, *World Water Vision* (The Hague, The Netherlands, March 2000).

26. B. R. Beattie and H. S. Foster Jr., "Can Prices Tame the Inflationary Tiger?" *Journal of the American Water Works Association,* 72 (August 1980): 444.

27. D. B. Gardner, "Water Pricing and Rent Seeking in California Agriculture," in *Water Rights: Scarce Resource Allocation, Bureaucracy, and the Environment,* ed. T. Anderson (Cambridge, MA: Ballinger Press, 1983), 83.

28. U. S. Geological Survey, *Estimated Use of Water in the United States in 1995,* Circular 1200 (Washington, DC, 1995).

29. Gleick, *The World's Water* (1998), 19.

30. World Health Organization, *Water Supply and Sanitation Sector Monitoring Report: Sector Status as of 1994,* rpt. WHO/EOS/96.15 (Geneva, 1996).

31. L. Nash, "Water Quality and Health," in *Water in Crisis: A Guide to the World's Fresh Water Resources,* ed. P. H. Gleick (New York: Oxford University Press, 1993).

32. International Institute for Applied Systems Analysis (IIASA), "Good to the Last Drop?" *Options* (summer 1996): 6.

33. Ibid.

34. U.S. Environmental Protection Agency and U.S. Department of Agriculture, *United States Clean Water Action Plan,* (Washington, DC, 1998).

CHAPTER 7. THE AIR WE BREATHE

1. World Health Organization, Press Release WHO/56, Meeting on Air Quality and Health, Geneva, September 14, 2000.

2. World Health Organization (WHO), *Guidelines for Air Quality* (Geneva: WHO, 2000).

3. Ibid., 42.

4. Mark Hertsgaard, *Earth Odyssey* (New York: Broadway Books, 1998).

5. Justino Regalado, "Air Pollution and Respiratory Health in Mexico City," *RT Magazine* (September 1, 2000).

6. Charles Dickens, *Hard Times* (1854; reprint London: Viking-Penguin, 1997).

7. U.S. Environmental Protection Agency (EPA), *National Air Pollutant Emission Trends, 1900–1996*, report EPA-454-R-97-011 (Research Triangle Park, NC: EPA, December 1997), fig. 3.3.

8. World Health Organization, *Guidelines for Air Quality*.

9. A. J. Haagen-Smit, "Chemistry and Physiology of Los Angeles Smog," *Industrial and Engineering Chemistry* 44 (1952): 1342; idem, "Abatement Strategy for Photochemical Smog," in *Photochemical Smog and Ozone Reactions*, Advances in Chemistry Series no. 113, ed. American Chemical Society (Washington, DC: American Chemical Society, 1972).

10. U.S. Environmental Protection Agency, *National Air Pollutant Emission Trends, 1900–1998*, report EPA-454-R-00-002 (Research Triangle Park, NC: EPA, March 2000).

11. The six principal ("criteria") air pollutants are sulfur dioxide (SO_2), nitrogen dioxide (NO_2), particulate matter (PM), ozone (O_3), carbon monoxide (CO), and lead (Pb).

12. U.S. Environmental Protection Agency, *Latest Findings on National Air Quality: 1999 Status and Trends*, report EPA-454/F-00-002 (Research Triangle Park, NC: USEPA, August 2000); U.S. Department of Energy, *Annual Energy Review* (Washington, DC: D.O.E. Energy Information Administration, 1999), table 7.3.

13. EPA, *National Air Pollutant Emission Trends*.

14. *PM-10* refers to all particles less than or equal to ten micrometers in diameter, which can penetrate deep into the lungs when inhaled.

15. This excludes agricultural dust and so-called "fugitive" dust from roads, construction sites, and mining operations.

16. S. Farrow and M. Toman, *Using Environmental Benefit-Cost Analysis to Improve Government Performance*, Resources for the Future, discussion paper 99-11 (Washington, DC: Resources for the Future, December 1998).

17. EPA, *Latest Findings on National Air Quality*.

18. T. O. Tengs et al., "Five Hundred Live-Saving Interventions and Their Cost-Effectiveness," *Risk Analysis* 15 (June 1995): 369.

19. D. W. Jorgenson and P. J. Wilcoxen, "Environmental Regulation and U.S. Economic Growth," *Rand Journal of Economics* 21(2) (summer 1990): 314.

20. J. C. Robinson, "The Impact of Environmental and Occupational Regulation on Productivity of United States Manufacturing," *Yale Journal on Regulation* 12 (summer, 1995): 387.

21. See, for example, C. S. Marxsen, "The Environmental Propaganda Agency," *The Independent Review* 5(1) (2000): 65.

22. Ibid., 69.

23. The American court system routinely places a monetary value on specific human lives in deciding cases involving, for example, car accidents, air-

plane crashes, and workplace casualties. Typical awards range from $400,000 for a death from a factory accident to $2.6 million for a student's death in the PanAm 103 bombing. In making awards, the trustees of the federal fund created to compensate families of the September 11, 2001, victims is considering factors such as financial need, future earning capacity, number of dependents, and psychological suffering. See D. B. Henriques, "In Death's Shadow, Valuing Each Life," *New York Times*, December 30, 2001, "Week in Review" sec.

24. M. Sagoff, *The Economy of the Earth* (Cambridge, UK: Cambridge University Press, 1988).

25. G. Likens, E. Wright, R. F. Galloway, and T. J. Butler, "Acid Rain," *Scientific American* 241 (1979): 43.

26. J. Harte, "Acid Rain," in *The Energy–Environment Connection*, ed. Jack M. Hollander (Washington, DC: Island Press, 1992); E. Cowling, "Acid Precipitation in Historical Context," *Environmental Science and Technology* 16 (1982): 110A.

27. A. H. Johnson and T. G. Siccama, "Acid Deposition and Forest Decline," *Environmental Science and Technology* 17 (1983): 294a.

28. National Acid Precipitation Assessment Program, *NAPAP Biennial Report to Congress: An Integrated Assessment* (Washington, DC: EPA, May 1998).

29. National Acid Precipitation Assessment Program, *Acidic Deposition: State of Science and Technology*, Final Report, ed. Patricia M. Irving (Washington, DC: U.S. Government Printing Office, 1991).

30. L. Roberts, "Learning from the Acid Rain Program," *Science* 251 (1991): 1302.

31. G. E. Likens, *The Ecosystem Approach: Its Use and Abuse*, Excellence in Ecology, ed. O. Kinne, bk. 3 (Oldendorf, Germany: Ecology Institute, 1992).

32. J. L. Kulp, "Acid Rain," in *The State of Humanity*, ed. J. Simon (Malden, MA: Blackwell, 1996), 523.

33. Although not new, this concern has been heightened in recent years by the increased globalization of trade and the evolution of international trade agreements such as the North American Free Trade Agreement (NAFTA), General Agreement on Tariffs and Trade (GATT), and the World Trade Organization (WTO), which replaced GATT in 1996. A serious rift exists between environmental groups, most of whom believe that uniform environmental standards should be included in all international trade agreements, and free trade advocates, who believe that forcing the environmental standards of the affluent onto poor countries would constitute a form of protectionism since the poorest countries, unable to comply with costly pollution measures, would find their exports to the affluent countries curtailed.

34. Interagency Task Force, *Review of U.S.–Mexico Environmental Issues* (Washington, DC: Office of the U.S. Trade Representative, October 1991), 194.

35. United States Census Bureau, *Pollution-Abatement Costs and Expenditures: 1994*, report MA200(94)-1 (Washington, DC, May 1996).

36. Organization for Economic Cooperation and Development (OECD), *The Effects of Government Environmental Policy on Costs and Competitiveness: Iron and Steel Sector,* report DSTI/SI/SC (97) 46 (Paris: OECD, 1997).

37. J. Tobey, "The Impact of Domestic Environmental Policies on Patterns of World Trade: An Empirical Test," *Kyklos* 43 (1990): 191.

38. R. Repetto, *Jobs, Competitiveness and Environmental Regulation: What Are the Real Issues?* (Washington, DC: World Resources Institute, May 1995).

39. Ibid.

40. H. Nordstrom and S. Vaughan, *Trade and Environment* (Geneva: World Trade Organization, 1999).

41. Ibid.

42. Glenn Martin, *San Francisco Chronicle* (December 30, 1997), 1.

CHAPTER 8. FOSSIL FUELS—CULPRIT OR GENIE?

1. Daniel Yergin, *The Prize* (New York: Simon & Schuster, 1991).

2. Ibid.

3. T. S. Wood and S. Baldwin, "Fuelwood and Charcoal Use in Developing Countries," *Annual Review of Energy,* vol. 10, ed. Jack M. Hollander (Palo Alto, CA: Annual Reviews: 1985), 407.

4. B. L. Turner II and Karl L. Butzer, "The Columbian Encounter and Land-Use Changes," *Environment* 34(8) (1992), quoted in R. A. Sedjo, "Forests: Conflicting Signals," *The True State of the Planet,* ed. R. Bailey (New York: Free Press, 1995), 182.

5. Sedjo, "Forests: Conflicting Signals."

6. D. W. MacCleery, *American Forests: A History of Resiliency and Recovery,* Forest Service report FS-540 (Washington, DC: U.S. Department of Agriculture [USDA], 1992)

7. D. Tillman, *Wood as an Energy Resource* (New York: Academic Press, 1978).

8. D.W. MacCleery, *What on Earth Have We Done to Our Forests?* Forest Service rpt., (Washington, DC: USDA, January 10, 1994).

9. D. S. Powell, J. L. Faulkner, D. R. Darr, Z. Zhu, and D. W. MacCleery, *Forest Resources of the United States, 1992,* Forest Service report RM-GTR-234 (Washington, DC: USDA, 1993); J. W. Barrett, ed., *Regional Silviculture of the United States* (New York: John Wiley, 1994).

10. T. S. Frieswyk, and A. M. Malley, *Forest Statistics for Vermont, 1973 and 1983,* Forest Service bulletin NE-87 (Washington, DC: USDA, 1985).

11. Powell et al., *Forest Resources of the United States.*

12. National Wilderness Preservation System, *Fact Sheet* (1994), NWPS Web site, www.wilderness.net/nwps/search.cfm.

13. Sedjo, "Forests: Conflicting Signals"; Powell, *Forest Resources of the United States.*

14. W. B. Smith, J. L. Faulkner, and D. S. Powell, *Forest Statistics of the United States, 1992*, Forest Service Report GTR_NC-168 (Washington, DC: USDA, 1994).

15. Sedjo, "Forests: Conflicting Signals."

16. MacCleery, *What on Earth Have We Done to Our Forests?*

17. Prominent in this effort is the Turning Point Project, a consortium of conservation groups.

18. R. A. Sedjo, *A Vision for the U.S. Forest Service: Goals for the Next Century* (Washington, DC: Resources for the Future, 2000).

19. United Nations Food and Agricultural Organization, *State of the World's Forests, 1997*, (Rome: FAO, 1997).

20. S. Vosti, *The Role of Agriculture in Saving the Rain Forest*, 2020 Vision Brief 9, International Food Policy Research Institute (February 1995).

21. *State of the World's Forests, 1997*. United Nations Food and Agricultural Organization, New York (1997).

22. Instituto Nacional de Pesquisas Espaciais, *Monitoring of the Brazilian Amazonian Forest by Satellite*, report CBERS-1 (São Jose dos Campos, Brazil, February 1999).

23. D. C. Nepstad, A. G. Moreira, and A. A. Alencar, *Flames in the Rain Forest: Origins, Impacts, and Alternatives to Amazonian Fires* (Brasilia, Brazil: World Bank, 1999).

24. Vosti, *Role of Agriculture in Saving the Rain Forest.*

25. "Managing the Rainforests," *Economist* (May 12, 2001), 83.

26. D. M. Wolfire, J. Brunner, and N. Sizer, *Forests and the Democratic Republic of Congo* (Washington, DC: World Resources Institute, 1998); J. Brunner, K. Talbott, and C. Elkin, *Logging Burma's Frontier Forests: Resources and the Regime* (Washington, DC: World Resources Institute, 1998).

27. D. Kaimowith and A. Angelsen, paper presented at conference in Costa Rica, March 1999, sponsored by the Center for International Forestry Research (CIFOR), Bogor, Indonesia.

28. W. F. Laurance et al. (eight co-authors), "The Future of the Brazilian Amazon," *Science* 291 (January 19, 2001): 438.

29. Carroll L. Wilson, *Coal—Bridge to the Future: Report of the World Coal Study* (Cambridge, MA: Ballinger, 1980).

30. N. Nakicenovic, "Freeing Energy from Carbon," *Daedalus* (special issue, ed. J. Ausubel) 125(3) (1996): 95.

31. Charles Dickens, *Hard Times* (1854; critical ed., New York: W. W. Norton, 1990).

32. Patrick E. Tyler, "China's Inevitable Dilemma: Coal Equals Growth," *New York Times*, November 29, 1995, A1.

33. Jack M. Hollander, "China and the New Asian Electricity Markets," *EPRI* [Electric Power Research Institute] *Journal* (September–October 1997): 25.

34. M. K. Hubbert, "Energy Resources," in *Resources and Man* ed. National Academy of Sciences–National Research Council (San Francisco: W. H. Freeman, 1969), 157.

35. R. A. Kerr, "The Next Oil Crisis Looms Large—and Perhaps Close," *Science* 281 (August 21, 1998): 1128.

36. U.S. Geological Survey, Fact Sheet 145-97 (Washington, DC, August 17, 1998).

37. A. B. Lovins et al., *Hypercars: Materials, Manufacturing, and Policy Implications* (Snowmass, CO: Rocky Mountain Institute, March 1996).

38. J. Ausubel, C. Marchetti, and P. S. Meyer, "Toward Green Mobility: The Evolution of Transport," *European Review* 6(2) (1998): 137.

39. Several electric-hybrid automobile models were being marketed in the United States in 2001, including the Honda Insight and the Toyota Prius. Ford has announced that it will market a hybrid version of the Ford Escape in 2003, while Daimler-Chrysler will produce a hybrid version of the Dodge Durango sport-utility vehicle.

40. U.S. National Academy of Sciences, *Energy in Transition, 1985–2010* (New York: W. H. Freeman, 1979).

41. Energy Information Administration (EIA), U.S. Department of Energy (Washington, DC, December 1998).

42. Ibid.

43. Gordon J. MacDonald, "The Future of Methane as an Energy Source," *Annual Review of Energy*, ed. Jack M. Hollander et al. (Palo Alto, CA: Annual Reviews, 1990), 53.

44. Washington Policy and Analysis, *Fueling the Future* (Washington, DC: American Gas Foundation, January 2000).

45. D. Jarvis et al., "Association of Respiratory Symptoms and Lung Function in Young Adults with Use of Domestic Gas Appliances," *The Lancet* 347 (1996): 426.

CHAPTER 9. SOLAR POWER TO THE PEOPLE

1. President Jimmy Carter, televised address to the nation (April 18, 1977).

2. Amory Lovins, "Soft Energy Technologies," in *Annual Review of Energy*, vol. 3, ed. Jack M. Hollander (Palo Alto, CA: Annual Reviews, 1978), 477.

3. U.S. Department of Energy, *Report of the President's Domestic Policy Review of Solar Energy* (Washington, DC, 1979).

4. Quoted in Robert Stobaugh and Daniel Yergin, *Energy Future* (New York, Random House, 1979), 183.

5. Dallas Burtraw, Joel Darmstadter, Karen Palmer, and James McVeigh, "Renewable Energy—Winner, Loser, or Innocent Victim?" *Resources* (spring 1999), 9.

6. Energy Information Administration, *Annual Energy Review* (Washington, DC: U.S. Department of Energy, 2000).

7. Ibid.

8. Energy Information Administration, *Annual Energy Review, 1997* (Washington, DC: U.S. Department of Energy, 1997).

9. Patrick A. March and Richard K. Fisher, "It's Not Easy Being Green," in *Annual Review of Energy and the Environment*, vol. 24, ed. Robert H. Socolow (Palo Alto, CA: Annual Reviews, 1999), 173.

10. Philip Fearnside, remarks made at meeting of the World Commission on Dams, São Paulo, August 1999.

11. J. S. Mattice, "Ecological Effects of Hydropower Facilities," in *Hydropower Engineering Handbook*, ed. J. S. Gulliver (New York: McGraw Hill, 1991).

12. Alliance to Save Energy, American Gas Association, and Solar Energy Industries Association, *An Alternative Energy Future* (Washington, DC, April 1992), 3:5; Robert L. Bradley Jr., *Renewable Energy: Not Cheap, Not Green*, Cato Policy Analysis no. 280 (Washington, DC: Cato Institute, August 1997).

13. March and Fisher, "It's Not Easy Being Green."

14. U.S. Department of Interior, U.S. Agency for International Development, *Hydropower's Environmental and Social Consequences, Including Potential for Reducing Greenhouse Gases* (paper presented at Kyoto Conference, Kyoto, Japan, November 1997).

15. Green Mountain Energy, Inc., *Your Guide to Renewable Energy Sources* (S. Burlington, VT, February 2000).

16. Water policy report (October 27, 1993), 30, quoted in Bradley, *Renewable Energy*.

17. Daniel P. Beard, remarks at the International Dam Summit, Nagaragawa, Japan, September 14, 1996.

18. Maria Gracinda Teixeira, *Energy Policy in Latin America* (Hants, UK: Ashgate Publishing, 1996).

19. American Wind Energy Association, *Global Wind Energy Market Report* (Washington, DC, 1999).

20. California Energy Commission (CEC), *Wind Energy in California*, CEC Web site at www.energy.ca.gov/wind/ (September 1, 1999).

21. National Audubon Society, *Audubon News* (November 3, 1999); John Bianchi, National Audubon Society, personal communication with the author, May 16, 2000.

22. Frank Harris and P. Navarro, *Policy Options for Promoting Wind Energy Development in California* (Graduate School of Management, University of California, Irvine, November 1999).

23. One should keep in mind the indirect government subsidies that the nuclear power industry continues to receive in the United States, especially the congressionally mandated limitation of liability in the case of accidents (Price-Anderson Act) and the continuation of large government-funded R&D programs that benefit nuclear power as well as basic science. These benefits will probably taper off in the future.

24. Bradley, *Renewable Energy*.

25. Harris and Navarro, *Policy Options for Promoting Wind Energy Development in California*.

26. Christopher Flavin and Nicholas Lenssen, *Power Surge: Guide to the Coming Energy Revolution* (New York: W. W. Norton, 1994).

27. Michael F. Northrop, Peter W. Riggs, and Frances A Raymond, *Solar: Financing Household Solar Energy in the Developing World* (report of a Rockefeller Brothers Fund workshop, Pocantico Hills, NY, October 11–13, 1995).

28. Ibid.

CHAPTER 10. NUKES TO THE RESCUE?

1. Data from International Atomic Energy Agency, *Newsbriefs*, Vienna (March 2000).

2. R. L. Garwin, *Can the World Do without Nuclear Power? Can the World Live with Nuclear Power?* (paper presented to Nuclear Control Institute, Washington, DC, April 9, 2001).

3. Bernard Cohen, *The Nuclear Energy Option: An Alternative for the '90s* (New York: Plenum Press, 1990). The caveat here is that the cost of extracting uranium from seawater today is at least fifty times the current market price of uranium. But this cost is likely to become competitive with terrestrial uranium in the next half-century.

4. Development of breeder reactors is controversial because they produce plutonium, which can be used either as a fuel for electricity production or as the explosive ingredient of nuclear weapons. See discussion of breeding later in the chapter.

5. P. Slovik, "Perceived Risk, Trust, and Democracy," *Risk Analysis* 13(6) (1993): 675; idem, "Perception of Risk," *Science* 236 (1987): 280.

6. This perception may be changing. In California an independent Field Institute poll in May 2001 indicated that 59 percent of Californians support building new nuclear power plants in the state, compared with 36 percent opposed and 5 percent undecided. This is a reversal of the results of the previous (1984) Field poll, in which only 32 percent of Californians favored new nuclear power. Also, a national survey sponsored by the Nuclear Energy Institute (an industry group) indicated softening in the public's view of nuclear power. To the statement "We should definitely build more nuclear energy plants in the future," 66 percent of respondents responded positively in March 2001 as compared with 42 percent in October 1999.

7. International Atomic Energy Agency, European Commission, and World Health Organization, *International Conference: One Decade after Chernobyl* (Vienna, April 8–12, 1996).

8. William C. Sailor, David Bodansky, Chaim Braun, Steve Fetter, and Bob van der Zwaan, "A Nuclear Solution to Climate Change," *Science* 288 (19 May 2000): 1177.

9. Spencer R. Weart, *Nuclear Fear* (Cambridge, MA: Harvard University Press, 1988).

10. Sailor et al., "A Nuclear Solution to Climate Change."

11. According to the UN Special Committee on the Effects of Atomic Radiation, the 528 atmospheric nuclear-weapons tests will be responsible over time for approximately three hundred thousand cancer deaths, while the Chernobyl accident will contribute some twenty-four thousand cancer deaths. For comparison, one year's normal operation of a single reactor has been estimated to cause about six cancer deaths from all components of the "fuel cycle," including mining, wastes, and reactor operations. See Garwin, *Can the World Live without Nuclear Power?*

12. R. A. Meserve, "What the National Energy Strategy Means for the Nuclear Power Industry" (speech presented to the Energy Investor Policy and Regulation Conference, New York City, December 4, 2001).

13. U.S. Nuclear Waste Technical Review Board, *Report to the U.S. Congress and the Secretary of Energy* (Washington, DC, 1999).

14. D. A. Ponce, *Interpretative Geophysical Fault Map across the Central Block of Yucca Mountain, Nevada*, U.S. Geological Survey open-file report 96-285 (Washington, DC, 1996).

15. John P. Holdren, *Improving U.S. Energy Security and Reducing Greenhouse-Gas Emissions: What Role for Nuclear Energy?* Testimony before Subcommittee on Energy and Environment, House Committee on Science, 106th Cong., 2d sess., July 25, 2000.

16. Ibid.

17. President Bill Clinton, "Non Proliferation and Export Control Policy," Presidential Decision Directive, September 27, 1993, *Federal Register* (January 21, 1997).

18. Committee on Future Nuclear Power Development, National Research Council, *Nuclear Power: Technical and Institutional Options for the Future* (Washington, DC: National Academy Press, 1992).

19. U.S. Nuclear Regulatory Commission, *Final Safety Evaluation Report Related to the Certification of the Advanced Boiling Water Reactor Design*, NUREG report no. 1503 (Washington, DC, 1994); Sailor et al., "A Nuclear Solution to Climate Change."

CHAPTER 11. WHEELS

1. American Automobile Manufacturers Association, *World Motor Vehicle Data* (Washington, DC, 1994).

2. U.S. Department of Energy, *International Energy Outlook, 1999: Transportation Energy Use*, report no. DOE/EIA-0484(99) (Washington, DC, 1999).

3. V. Wouk, "Hybrid Electric Vehicles," *Scientific American* (October 1997): 70–74.

4. This is in contrast to the situation with solar energy technologies, which face a fundamental obstacle to low cost owing to the diluteness of the sun's energy reaching the earth's surface. See Chapter 9 for details.

5. Texas Transportation Institute, *The 1999 Annual Mobility Report: Information for Urban America* (College Station, TX, 1999).

6. D. Shrank, S. Turner, and T. Lomax, *Estimates of Urban Roadway Congestion, 1990*, report 1131-5 (College Station, TX: Texas Transportation Institute, 1993).

7. Apogee Research, Inc., *The Road Information Program National Transportation Survey: 1990 Poll Results* (Washington, DC: Road Information Program, 1990).

8. Bureau of Transportation Statistics, *Highway Statistics 2002*, (Washington, DC: U.S. Department of Transportation, 2002). table HM-20.

9. Ibid., figure 2-12.

10. Ibid.

11. P. Varaiya, "Making Freeways Work," *Access*, no. 16 (spring 2000): 22.

12. R. Gaurav and C. J. Khisty, *Urban Transportation in Developing Countries: Trends, Impacts, and Potential Systemic Strategies* (paper presented at 77th Annual Meeting, Transportation Research Board, Washington, DC, January 1998).

13. R. K. Bose, "Automobiles and Environmental Sustainability: Issues and Options for Developing Countries," *Asian Transport Journal* (December 1998): 13.1–13.16.

14. P. Midgley, *Urban Transport in Asia: An Operational Agenda for the 1990s* (Washington, DC: World Bank, 1994).

15. Bose, "Automobiles and Environmental Sustainability."

16. W. Owen, "Global Transportation," *Access*, no. 13 (Fall 1998).

17. W. Owen, *Transportation and World Development* (Baltimore: Johns Hopkins University Press, 1987).

CHAPTER 12. DON'T HARM THE PATIENT

1. H. von Staden, trans., Hippocratic Oath, *Journal of the History of Medicine and Allied Sciences* 51 (1996): 406.

2. The evocative title *Patient Earth* was used by J. Harte and R. Socolow for the book they edited, describing some environmental case histories, including the story of Tennessee's Tellico Dam and an endangered minnow (New York: Holt, Rinehart and Winston, 1971).

3. World Wildlife Fund, full-page advertisement in *New York Times*, August 21, 1998.

4. L. S. Kaufman, L. J. Chapman, and C. A. Chapman, "Evolution in Fast Forward: Haplochromine Fishes of the Lake Victoria Basin," *Endeavour* (Cambridge, UK) 21(1) (1997), 23.

5. O. Seehausen, F. Witte, E. F. Katunzi, J. Smits, and N. Bouton, "Patterns of the Remnant Chiclid Fauna in Southern Lake Victoria," *Conservation Biology* 11(4) (1997): 890.

6. P. Ehrlich, "Intervening in Evolution: Ethics and Actions," *Proceedings of the National Academy of Sciences* 98 (10) (May 8, 2001): 5477.

7. See, for example, J. L. Simon, and A. Wildowsky, "Species Loss Revisited," in *The State of Humanity,* ed. J. L. Simon (Malden, MA: Blackwell, 1995).

8. Biologist N. Myers observes that "with the current rates of exploration and extinction we shall never have more than best-guess approximations" of these quantities. See Myers, review of *A Convincing Call for Conservation, Science* 295 (January 18, 2002): 447.

9. Bjorn Lomborg, *The Skeptical Environmentalist* (Cambridge, UK: Cambridge University Press, 2001).

10. A. N. James, *National Investment in Biodiversity Conservation: A Global Survey of Parks and Protected Areas Agencies* (Cambridge, UK: World Conservation Monitoring Centre, April 1996).

11. Responsibility lies with the Secretary of the Interior, through the U.S. Fish and Wildlife Service and, for marine species, the National Marine Fisheries Service.

12. U.S. Fish and Wildlife Service, *Endangered Species General Statistics;* Website at www.fws.gov/~r9endspp/esastats.html (1997).

13. W. Beckerman, *Through Green-Colored Glasses,* (Washington, DC: Cato Institute, 1996), 85.

14. Paul Rauber, "The Great Green Hope," *Sierra Magazine* (July–August 1997).

15. National Wilderness Institute (NWI), "Endangered Species Blueprint," *NWI Resource* 5(1) (fall 1994): 1.

16. I. Sugg, "If a Grizzly Attacks, Drop Your Gun," *Wall Street Journal,* November 13, 1993, A15.

17. Simmons, R.T., "The Endangered Species Act: Who's Saving What?" *The Independent Review* 3(3) (winter 1999): 309–326.

18. U.S. Fish and Wildlife Service, *Report to Congress on the Endangered and Threatened Species Recovery Program* (Washington, DC: U.S. Government Printing Office, 1993).

19. C. M., Wilkinson, in a paper quoted in Simmons, "The Endangered Species Act"; J. Adler, testimony before the Senate Committee on Environment and Public Works, 105th Cong, 2d sess., July 12, 1995.

20. Helen Chenoweth-Hage, (R-Idaho), testimony before the House Resources Committee on reauthorization of the Endangered Species Act, 106th Cong., 2d sess., February 2, 2000.

21. W. J. Snape II and R. M. Ferris, *Saving America's Wildlife: Renewing the Endangered Species Act,* report no. 4 on the Endangered Species Act, published on Defenders of Wildlife Web site, www.defenders.org (2001).

22. M. McCabe, "Gray Wolves Heading to California," *San Francisco Chronicle,* February 5, 2002, A-1.

23. E. O. Wilson, *The Future of Life* (New York: Knopf, 2002).

24. C. C. Mann and M. L. Plummer, "A Species' Fate, by the Numbers," *Science* 284 (April 2, 1999): 36.

25. R. Carson, *Silent Spring* (New York: Houghton-Mifflin, 1962).

26. B. Ames, "The Causes and Prevention of Cancer," in *The True State of the Planet*, ed. R. Bailey (New York: Free Press, 1995).

CHAPTER 13. CHOICES

1. World Commission on Environment and Development (the "Brundtland Commission"), *Our Common Future* (Oxford, UK: Oxford University Press, 1987).

2. Vandana Shiva, *Poverty and Globalization* (Reith Lecture, British Broadcasting Corporation, London, May 10, 2000).

3. World Commission on Dams, *Dams and Development: A New Framework for Decision-Making* (London: Earthscan Press, November 16, 2000).

4. Ibid.

5. World Bank and International Monetary Fund, *Review of the PRSP Experience* (staff paper) (Washington, DC, January 7, 2002).

6. Ibid.

7. Byron G. Auguste, "What's So New about Globalization?" *New Perspectives Quarterly* (January 1, 1998).

8. Ibid.

9. United Nations Development Programme, *Human Development Report, 2001* (Oxford, UK: Oxford University Press, 2001).

10. Ibid.

11. Hernando de Soto, *The Mystery of Capital: Why Capitalism Triumphs in the West and Fails Everywhere Else* (New York: Basic Books, 2000).

12. J. Madrick, "The Charms of Property," *New York Review of Books* (May 31, 2001), 39.

13. Ibid.

14. United Nations, *Overcoming Human Poverty*, Poverty Report 2000 (New York: United Nations Development Programme, 2000).

15. United Nations data quoted by Clare Short, Member of Parliament, speaking at the Rockefeller Foundation, New York, February 1, 2001.

16. Amartya Sen, *Development as Freedom* (New York: Knopf, 1999).

INDEX

ABOUT THE AUTHOR

Jack M. Hollander is a professor emeritus of energy and resources at the University of California, Berkeley. Author or coauthor of over one hundred research publications and editor of twenty books, Dr. Hollander has undertaken basic research in nuclear structure physics, energy and environment research, academic administration, and science and technology development. A 1951 UC Berkeley Ph.D. in chemistry, he was one of Berkeley's early researchers in the environmental sciences in the 1960s and first director of both the Berkeley Laboratory's Energy and Environment Division and the systemwide University of California Energy Institute. In the mid-1970s he served as director of the first (and only) national energy study carried out by the U.S. National Academy of Sciences and for seventeen years was editor of the independent book series Annual Review of Energy and the Environment. Dr. Hollander served for twelve years as chairperson of the Swedish Academy's International Institute of Energy and Human Ecology, based in Stockholm. He was also cofounder of the independent American Council for an Energy-Efficient Economy (ACEEE). During the 1980s Dr. Hollander served as vice president for research and graduate Studies at The Ohio State University, Columbus. He was the recipient of two Guggenheim fellowships, in 1958 and 1966.

Compositor: Michael Bass & Associates
Text: 10/13 Aldus
Display: Univers Condensed
Printer and binder: Maple-Vail Manufacturing Group